Robert Boyle and Son

Natural and Artificial Methods of Ventilation

Robert Boyle and Son

Natural and Artificial Methods of Ventilation

ISBN/EAN: 9783337030476

Printed in Europe, USA, Canada, Australia, Japan

Cover: Foto ©berggeist007 / pixelio.de

More available books at **www.hansebooks.com**

Natural and Artificial

Methods of

Ventilation.

"*A People's Health's a Nation's Wealth.*"—FRANKLIN.

LONDON:

ROBERT BOYLE & SON, LIMITED,

64, HOLBORN VIADUCT.

AND AT

GLASGOW, PARIS, BERLIN, AND NEW YORK.

1899.

"*I always find theoretical men rather inclined to look with a certain amount of disdain upon practical men, and to think that practical men know nothing about their subject unless they follow the dictates of theorists. But they are now paying much more attention to the teachings of experience.*"—

SIR WILLIAM PREECE, F.R.S.

Introduction.

THIS compilation is published with the view
of demonstrating the comparative values of
natural and artificial methods of ventilation, when
the former is scientifically applied, and how
ventilation may be successfully achieved with the
simplest means by an intelligent comprehension of
the laws which govern the movements of air and
the utilisation of the powerful natural forces which
are unceasingly in operation.

The action of natural and of mechanical methods
of ventilation is described and illustrated by diagrams,
and an exposition given of the opinions held by the
accepted authorities on the subject, with excerpts from
the published accounts of their personal experience
of the different systems.

Extracts are also given from the official reports
of Commissions on ventilation, appointed by the
British, French, and American Governments, in which
the evils of forced downdraught ventilation and of
hot-air heating are described, and the dangers of
open-window ventilation in cold weather.

The evidence submitted in the following pages
in favour of or against the respective systems emanates
from well-known Scientists and Sanitarians whose
names are a sufficient guarantee of the value of the
statements they make and of the opinions they
express. Theories are not dealt with, the results of
practical work only being recorded.

R. B. & S.

"*Air is the prime supporter of life; health; even life itself is dependent upon its purity.*

"*Statistical inquiries on mortality prove beyond a doubt that of the causes of death which are usually in action, impurity of the air is the most important.*"

—PARKES.

Ventilation.

" Ventilation is a science, and it requires the study of a lifetime to master properly all its intricacies. The greatest engineering skill is necessary in the arrangement of tubes and the supply of fresh air."—PARKES.

There is perhaps no other subject with respect to which there is greater diversity of opinion than that of ventilation. Ventilation is the most difficult of all sciences to practically deal with in such a manner as to satisfy every one. Indeed to do so is well-nigh impossible, owing to the varying idiosyncrasies and temperaments of different individuals. It will, however, be admitted that the method which secures the requisite change of air in the simplest and least objectionable manner, is the one most likely to prove generally acceptable.

That that method is the natural one when properly applied is conceded by the highest authorities on the subject. It possesses the advantage of being easily installed at a moderate cost, and, requiring no special attention, it can never get out of order or break down as so often happens with mechanical ventilation.

It is also admitted that the proper place from whence to extract the warm expired air is at the ceiling, or upper part of a building, to where it naturally ascends.

Dr. Parkes, than whom there is no higher authority, says: "As the ascent of respired air is rapid, on account not only of its temperature, but from the force with which it is propelled upwards, the point of discharge should be above.

"By some it has been argued that it is better that the foul air should pass off below the level of the person, so that the products of respiration may be immediately drawn below the mouth, and be replaced by descending pure air. But the resistance to be overcome in drawing down the hot air of respiration is so great that there is a considerable waste of power, and the obstacle to the discharge is sometimes sufficient, if the extracting power be at all lessened, to reverse the movement.

"This plan in fact must be considered a mistake. The true principle is that stated long ago by D'Arcet: 'In the case of vapours or gases the proper place of discharge is above.'

"Three forces act in natural ventilation, viz., diffusion, wind, and the difference in weight of masses of air of unequal temperature.

"In temperate climates in most cases natural ventilation is the best. Incessant movement of the air is a law of nature. We have only to allow the air in our cities and dwellings to take share in this constant change, and ventilation will go on uninterruptedly without our care.

"The evidences of injury to health from impure air are found in a larger proportion of ill-health—i.e., of days lost from sickness in the year—than under any other circumstances; an increase in the severity of many diseases, which, though not caused, are influenced by impure air, and a higher rate of mortality."

Another authority says: "Anything which has passed through the human body ought to be treated as excreta and rejected;—just as sewage was thrown away into the drains, so air that had passed through the human lungs should be got rid of at the earliest possible moment without allowing it to go through the lungs of someone else. It was as unreasonable to breathe the same air twice, or twelve or twenty times over, as was the case in many places, as it would be to go to the sewer for drinking water."

The old-fashioned formula which used to be employed to determine the amount of ventilation necessary for a given number of people by the height and area of a shaft or flue is now obsolete, the calculation being based upon the supposition that a column of air equal to the area and length of the shaft would always pass up and out; but it has been recognised that this is a fallacy, as at times the air passed down as freely as it passed up, and that occasionally no current either way existed at all.

Atmospheric air per 1,000 volumes is composed of oxygen 209·6, nitrogen 790·0, carbonic acid 0·4. Expired air contains about 5 per cent. less of oxygen, and a little more than 4 per cent. more of carbonic acid than that which is inhaled. From 25 to 40 ounces of water is passed off from the lungs and skin in twenty-four hours.

Carbonic acid gas is not in itself poisonous, but as a product of respiration, particularly in connection with an excess of moisture, its presence is always an index of contamination of the air by other impurities, such as organic matter, and it is therefore employed as the standard to determine the vitiation.

Carbonic acid diffuses in the air and does not gravitate to the lower levels.

One cubic foot of gas consumes the oxygen of about 8 cubic feet of air and produces 2 cubic feet of carbon dioxide.

The average number of respirations made by an adult is 20 per minute, 30 cubic inches of air being inhaled at each respiration, or 600 cubic inches per minute, or about 20 cubic feet per hour.

With a temperature of 70° F. the temperature of the air expelled from the lungs is from 85° to 95° F.

It is the organic matter suspended in the watery vapour expelled from the lungs and exhaled from the body wherein the real danger lies. As this vapour is in a heated state and immediately ascends, it should be drawn off at the highest point and not permitted to return to be rebreathed as happens with downward ventilation by propulsion, a method which has justly been described as "a standing menace to the health of society."

With mechanical ventilation the air is propelled or drawn in at a high velocity ranging from 5 feet to 20 feet per second, destructive of diffusion, and fatal to the effective or comfortable ventilation of a building. As one authority says, "Engine-driving columns of air through a building is not ventilating it."

With downward ventilation by propulsion from three to four times the volume of air is required that is necessary with upward ventilation, the cost being proportionately increased.

Extraction by fans has not been more successful, and they have in many cases to be discarded owing to the unbearable draughts they create.

Hot-air aspirating shafts are a costly and unsatisfactory method of ventilation, and are now seldom employed.

Ventilation by dilution necessitates the supply of a very large volume of air to keep the air in a building in anything like a healthy state. When the ventilation is by extraction the vitiated air is removed as fast as it is generated, and healthy ventilation is secured, with but a fraction of the volume of air required for dilution by propulsion.

One of the greatest dangers to health is where the fresh air supply is raised to a temperature such as is required to effectively heat a building, as it is thereby seriously deteriorated and rendered unfit for healthy breathing purposes.

The heating of a building should always be kept separate and distinct from that of the ventilation. Radiant heat is the healthiest.

The velocity of the air supply should never be greater than 2 feet per second, if draughts are to be avoided. The fresh air should be admitted at about breathing level in an upward direction through a number of small tubes distributed round the room to secure the most perfect diffusion. Where the air is warmed by passing through radiators a lower level is permissible.

The great majority of the bacteria found in air are not only perfectly harmless but are beneficial to human beings, they acting as so many scavengers of the air, and to get entirely rid of them would be prejudicial to health.

6

Volume of Air required for Ventilation.

" As the authorities on the subject hold such widely divergent views as to the amount of air required for healthy ventilation, and there is no accepted standard, each system must of necessity be adopted on its own merits."—VENTILATION.

"With the many diverse methods of ventilation which are at present in use there can be no fixed standard for determining the volume of air necessary to secure effective ventilation; one system ensuring with a given volume a healthy atmosphere, to secure which with another system several times that volume may be required.

"At a time when the science of ventilation, and the natural laws which govern it, were but imperfectly understood, it was held by many, and still is by some, that a very large volume of air was necessary to *dilute* the vitiated air in a building to a healthy standard, from 2,000 to 6,000 cubic feet per hour per person being recommended. Experience, and a more enlightened knowledge of the subject, have, however, taught us that with the more perfect forms of ventilation by *extraction* from the upper parts of a building, to where the warm expired air naturally ascends, considerably less than the least of these estimates is sufficient for all ordinary requirements. A high authority fixes 100 cubic feet per hour per person as quite sufficient with this method of ventilation, as the average of 20 cubic feet of air which is respired by an adult per hour is expelled from the lungs at a temperature of 80° to 100° F., and ascends, along with the heated exhalations from the body, to the upper parts of the building, and is there at once drawn off, so that no part of the vitiated air can be returned to be rebreathed, as is the case when the ventilation is effected by dilution by impulsion, particularly when on the downward principle.

"As with effective upward ventilation the ascending vitiated air is drawn off as fast as it is generated, an equivalent supply of fresh air being admitted at or below breathing level, it is obvious that a considerably less volume of air would be required with this method to secure efficient ventilation than where the foul air is merely *diluted* to a given standard, as is the case with all forced systems of ventilation by impulsion, necessitating the introduction of large volumes of air."

Professor R. H. Smith says: "The commonly adopted basis of calculation of so many cubic feet of space in each room per person meant that the object aimed at was the slowing down to a standard time-rate of the vitiation of a stationary quantity of air. From this idea was derived that of supplying per hour between 30 and 200 times as much as was actually inhaled by the inmates of the room. The true idea of perfect ventilation is evidently to inject and extract only a moderate excess, say five to ten times as much, over that actually used, and to do so in such a manner that (1) the exhalations do not mix with the fresh air supply, and (2) the inflow is properly diffused and does not pass direct to the outlets in merely local currents or draughts."

8

With downward ventilation by propulsion from three to four times the volume of air is required that is necessary with upward ventilation.

Professor Smith, in a recent article in the *Engineer*, states, "In order to keep down the percentage of pollution to a non-dangerous degree, under this system [forced downdraught] arises, therefore, the necessity of admitting for ventilation fresh air in quantities many times greater than that actually used, and also a correspondingly extravagant expenditure of heat if this supply be artificially warmed. Thus the only ideally perfect ventilation consists in inducing a regular up-current from a level below that of the human head up to the extraction outlets at the ceiling. Under this system the bulk of fresh air required to be admitted is immensely reduced, as is also the expense of warming it to any degree considered desirable."

Velocity of the Air Supply.

"Engine-driving columns of air through a building is not ventilating it"

To secure the most perfect diffusion, equable ventilation, and freedom from draught, the fresh air supply should be admitted at a *low* velocity. The highest authorities fix the maximum speed at from 1½ feet to 2 feet per second. The air should be admitted directly through the walls into the room at a low level and in an upward direction.

The *Builder* (April 30th, 1898), referring to a mechanical system of ventilation, says :—" Mr. . . . lays it down as an axiom that the rate of air motion through the air inlets into the room should not be more than 5 feet per second. This is, in our opinion and in that of good authorities, too much ; and 2 feet per second is far preferable."

How a Natural System of Ventilation Acts.

" As the warm air expands it ascends, . . . as the warm air ascends the air around rushes in to fill its place."—SIR DOUGLAS GALTON, F.R.S.

A scientifically arranged natural system of ventilation utilises the never-ceasing movement or natural force which exists in the atmosphere as an unfailing motive power, in conjunction with the difference in temperature of the internal and external atmosphere.

As applied to buildings, it consists of specially constructed updraught ventilators which remove the pressure of the external air from the top of the outlet shafts and create, under all conditions of the weather, a continuous and powerful exhaust at the higher parts of the building; combined with properly arranged air-inlets, fixed at the lower levels, admitting the air directly through the walls in an upward direction at a low velocity, purified, and warmed or cooled as required, ensuring a constant change of air and perfect diffusion of the fresh air supply in accordance with the natural laws which govern ventilation.

The higher temperature in an occupied building

B

—particularly in cold weather, when doors and windows are kept closed to exclude draughts—combined with the ascensional movement of the heated air arising from the lungs and body, are also utilised as powerful and never-failing auxiliaries in changing the air and inducing an inward and upward current, even when there is absolutely no *perceptible* movement of the air outside.

"Incessant movement of the air is a law of Nature. We have only to allow the air in our cities and dwellings to take share in this constant change, and ventilation will go on uninterruptedly without our care."—PARKES.

Why a Natural System of Ventilation is Always in Operation.

" *Science proves that there is not a moment of time but when there is a movement of the air, and that this movement properly utilised is sufficient at all times to change the air in a building and secure ventilation.*"—HOUGHTON.

A scientifically arranged system of natural ventilation may be described as the application of means by which the natural laws of ventilation can be effectively brought into operation, the power which operates in producing the currents and change of air being as constant as gravity.

"Incessant movement of the air is a law of Nature; we have only to allow the air in our cities and dwellings to take share in this constant change, and ventilation will go on uninterruptedly without our care.

In this country, and, indeed, in most countries, even
comparative quiescence of the air for more than a few
hours is scarcely known. Air is called 'still' when it
is really moving 1 or 1½ miles an hour. Advantage,
therefore, can be taken of this aspirating power of the
wind to cause a movement of the air up a tube."—
PARKES.

"That efficient ventilation can ever be automatic
and costless may, perhaps, appear absurd; it is, however, not so absurd as it may appear. This will be
evident by reference to the natural laws of atmospheric pressure and of its expansion by heat. In fact,
the idea that the atmosphere cannot be caused to pass
through a building at a rate of emptying and replenish-
ing every twenty minutes, by having the inside warm
and the outside cold, with sufficient openings at the
bottom and top, is itself absurd to anyone familiar
with dynamics. . . The outside air presses in through
the inlets and the inside air presses out through the
outlets at a speed determined by the difference in
weight; that is, in winter, a difference between air
averaging about 35° F., or lower, and about 65° F., or
higher.

"Now, according to the laws of atmospheric
pressure, this difference produces a speed of at least
8 feet per second, in ordinary flues or chimneys. (See
Blue Book, p. 6.)

"'The ventilation of rooms and buildings,' say
the Government Commissioners, 'depends in a great
measure upon the ascending and descending currents
of air caused by the difference in temperature, the
warmer air ascending and the colder and heavier
descending; the greater the difference of temperature

the greater being the difference of weight and of the rapidity of interchange. . . A cubic foot of air—

at 30° F., weighs   ...   569·2 grains
at 70° F., weighs   ...   526·2 grains
_____
making a difference of ...   43·0 grains

for the 40° under the same atmospheric pressure, viz., 30 inches of mercury.' (Blue Book, pp. 5, 6.) This difference in weight produces a speed of at least 10 feet per second, in ordinary flues or chimneys.

"If, therefore, the inlets and outlets be properly proportioned and open, the ordinary atmospheric pressure will carry on the ventilation quite efficiently, and the whole hospital will be kept fresh and comfortable by the natural forces alone. There is no fear that the speed will not be enough to keep up efficient ventilation—it is more likely to be too great; but there need be no fear of its being too great, because it is completely under control, and can be regulated to any rate desired by the valves at the ward inlets and outlets. Natural ventilation is certainly much to be preferred to any and every artificial system, whether on the 'Plenum' or vacuum principle, and it is, of course, much less complicated. It is, indeed, comparative simplicity itself—merely arranging openings and warming the air, and, as shown above, it acts more than efficiently. It also involves very little original outlay, and comparatively no permanent cost for maintenance. Whereas all artificial systems involve costly original plant of machinery, as well as heavy permanent expense for maintenance in engines, engineers, fuel, &c., and with all they cannot be made as efficient or nearly so pleasant and healthy in operation."—Dr. John Hayward.

Advantages of a Natural System of Ventilation.

" As the ascent of respired air is rapid, on account not only of its temperature, but from the force with which it is propelled upwards, the point of discharge should be above."—PARKES.

With a natural system of ventilation the warm vitiated air is continuously drawn off, as fast as it is generated, at the highest point to where it naturally ascends, an equivalent supply of fresh air entering at the lower levels through specially arranged inlets.

Efficiency is secured in every season of the year, and in all conditions of the weather—in the coldest and foggiest day in winter and the closest and warmest day in summer.

The fresh air supply is admitted at a *low* velocity directly through the walls in an upward direction and is thoroughly purified, and warmed or cooled as required.

The special construction of the air inlet openings, gratings and tubes completely prevents the lodgment or accumulation of dust. They are also easily accessible for cleansing purposes. The importance of these features cannot be overrated.

The air in a building is kept perfectly clear and pure even when there is a dense fog outside. It is always in action, day and night, and can never get out of order or break down, there being no movable parts.

Requires the minimum of attention and is under perfect control.

Can be installed at a fraction of the cost of mechanical or other artificial methods, and there is little, or no, after-expense for maintenance, according to the appliances adopted.

When Natural Ventilation Fails.

" It is essential to the success of a natural system of ventilation that both the outlet and inlet ventilators be of correct construction and skilfully applied. Where this is not observed failure generally ensues with this form of ventilation."—DE CHAUMONT.

Natural ventilation generally fails—

(1) When, owing to a mistaken spirit of economy, a *complete* system is not adopted.

(2) Where the outlet and inlet ventilators are insufficient in size and number, and not properly proportioned to each other.

(3) Where the ventilators are of defective make, or are unskilfully arranged in their relation to each other.

" My experience is that a natural system of ventilation is, in the long run, the most reliable and satisfactory, provided, of course, that properly constructed ventilators are employed, and that they are skilfully applied by a competent engineer. I am quite aware that there are many very indifferent ventilators in existence, and also incompetent persons who style themselves 'ventilating engineers,' and that natural ventilation under such auspices might, and probably would, prove a failure."—HOUGHTON

" The reputation of natural ventilation has suffered a good deal from the abortive attempts of individuals having little or no acquaintance with either the science or the practice of ventilation, and by the employment of so-called ventilating apparatus of crude and unscientific construction. The selection of ventilators is likewise sometimes left by an architect to the contractor, who naturally supplies those upon which he gets the largest profit, without troubling himself about their efficiency."—*Building News.*

How to Effectively Ventilate a Building with a Natural System of Ventilation.

———

"*We have always maintained that a system of ventilation which could be universally applied must be of such a nature that it cannot get out of order, is independent of any special attention, and is self-acting in every part.*"—ENGINEERING.

———

Natural ventilation properly applied is the most constant and reliable. It requires little or no attention, is always in action, and cannot get out of order or break down, as so often happens in mechanical ventilation.

Care should be taken in the employment of ventilators that they are of correct construction, and that both the outlets and inlets are sufficient in size and number, and so placed as to effectively accomplish the work they have to do.

The outlet ventilators should be fixed on the highest part of the roof, clear of all obstructions, so that the wind can reach them freely from every quarter.

The fresh air supply should be properly proportioned to the extraction.

The combined area of the inlets should, in all ordinary cases, be at *least* equal to that of the outlet shafts.

Main exhaust shafts should be at least of equal area to the combined area of the branch pipes.

Branch pipes should have as great an upward angle as possible, and should never enter the main shaft at the same level unless when parallel.

All pipes should be made of metal, and circular in shape to reduce friction. Wood shrinks, and is not air-tight.

The vitiated air should be extracted at the ceiling, to where it naturally ascends.

The fresh air should be admitted directly through the walls at a low velocity, in an upward direction, through a number of small inlet brackets or tubes distributed round the walls to secure more complete diffusion and an equable movement of the air in all parts of the building.

Air should never be admitted in cold weather in a horizontal direction, as a disagreeable draught would result.

The velocity of the air supply should not exceed 2 feet per second.

The inlet channels should communicate directly with the outer air, be as short as possible and easy of access for the purpose of cleaning.

Long inlet channels are objectionable, as they harbour dirt and are difficult of access.

Where simple inlet tubes or brackets are used, the tops should be about 5 ft. 9 in. above the floor.

Where the air supply is warmed it may be delivered at a lower level.

The inlets should be fitted with regulating valves to control the air supply.

The fresh air supply should not be overheated, as its hygienic properties are thereby seriously impaired. Heating a building by hot air should, therefore, be avoided as injurious to health.

Radiant heat is the healthiest and most effective.

Where open fires are used, they should have a separate air-supply, led directly to them from the outside air through a 3-in. or 4-in. pipe, to prevent them drawing the hot vitiated air from above down into the breathing zone, creating draughts, and unduly affecting the ventilating arrangements.

Open Window Ventilation in Cold Weather.

"Any method of ventilation which permits of a downdraught of cold air is injurious to health."—DE CHAUMONT.

When a window is open at the top in cool weather, a stream of cold air passes into the room, spreads, and descends in a cold shower, cooling and pressing down the warm ascending vitiated air to be rebreathed, along with the poisonous products of combustion, and the dust and dirt with which the incoming air may be charged.

It is in cold weather that special ventilating arrangements, that will change the air without draught and at the same time purify it, are absolutely necessary. In warm weather, windows can safely be kept open, there being then not the same danger from draughts.

"For winter, and for spring and autumn, the outer air in our climate is generally too cold to be admitted through open windows. It is, of course, impossible to warm the air as it enters, before it impinges on the patients, when it is let in through open windows and the warming is done by open fires, because the patients lie between its entrance and the means of warming.

"Letting the air in through the windows tends to drive the air of the wards into the hospital, which is always objectionable; and is also risky when there is infectious disease in the wards.

"In a Blue Book, issued in 1898, 'On the Ventilation and Warming of Board Schools,' on p. 23 occurs this statement in reference to ventilation by open windows: 'In those cases in which the ventilation was effective the temperature of the dormitory would follow closely that of the outside air, and would not exceed that of the outside air by more than about 2½"'. . .

"In spring and autumn the weather is then generally too cold to have the windows always open and yet not cold enough to have the fires continually burning, so that the abstraction cannot then be accomplished efficiently by either windows or fires; it must consequently be done by outlets specially provided for the purpose.

"I am aware that some medical practitioners do recommend hospitals to be ventilated only by open windows on both sides of large wards, even in winter in this climate, and maintain that there is neither risk nor discomfort in doing so. They do not themselves, however, submit to such ventilation . that is, they do not themselves sit in their study, or have their families to sit in the morning room or sleep in bedrooms with windows to the east open all the winter through! And it is only bounce for matrons and nurses to acquiesce in such an arrangement and uphold it; they are themselves not confined to beds underneath open windows, but are moving about in and out of the wards. Nor do they submit to it in their own rooms during cold easterly windy days and nights.

"The helpless patients scarcely dare complain or even admit that they feel a draught if the nurse says they do not."—DR. JOHN HAYWARD.

Open Fire Ventilation.

" The air of a room may be changed by this method, but at a sacrifice of comfort, and consequent injury to health."—RICHARDSON.

Though an open fire undoubtedly acts as a powerful exhaust, it is well known to sanitarians that no more dangerous mode of changing the air in a room exists, owing to the currents of cold air set up towards it from doors and windows, causing disagreeable draughts to be experienced by those in its vicinity, or in any part of the room which the currents traverse.

The upper strata of air in the room, charged with the heated products of respiration and combustion, is drawn downwards towards the fire and inhaled by the occupants, who are thus compelled to continuously rebreathe exhausted and impure air.

" For hospital wards open fires are risky, costly, and troublesome ; they greatly embarrass all plans of ventilation, and are in many other ways objectionable. They are of but little use as means of ventilation ; they certainly abstract air, but they take it principally from near the floor, whereas the foul air accumulates near the ceiling, being heated and so made light by the same means that foul it. Hospitals warmed only by open fires are sure to be very imperfectly warmed ; and if at the same time they are attempted to be ventilated only by open windows they are most certainly also very imperfectly ventilated. It is much to be regretted that some hospitals are professed to be ventilated and warmed by simply open windows and open fires, even throughout the winter in this climate."—DR. JOHN HAYWARD.

Open Pipe Ventilation.

It is almost invariably found that where plain open pipes are used, or have merely a cap on the top to exclude rain, they are stopped up on account of the downdraught experienced.

"Gusts of wind will sometimes cause a reverse action in the tube. In hot weather if the shaft is colder than the outer air a down-current may ensue.

"In consequence of the numerous causes of disturbance enumerated above, this method of extraction, when applied to a house, could not be relied on to act on all occasions with certainty as an extraction shaft."—SIR DOUGLAS GALTON, F.R.S.

Revolving Ventilators.

Revolving ventilators soon get out of order, particularly if employed for the ventilation of drains or sewers, as the gases quickly corrode the working parts, rendering the ventilator worse than useless, owing to the down current which results in consequence of its ceasing to revolve with the required rapidity.

"The objection to a movable ventilator is that if by any accident it failed to turn it would become a powerful inlet."—SIR DOUGLAS GALTON, F.R.S.

Louvre Board Ventilators.

"Louvre Board ventilators were originally designed for the purpose of preventing the weather from passing into a shaft. The large body of air which it admits at the top of the shaft, and which is supposed to pass across from one side to the other and thence out, simply acts as a damper upon the ascending current, and, in cold weather particularly passes down the shaft precipitating the vitiated air and causing a disagreeable downdraught."—VENTILATION.

Air Analysis.

"In buildings, such as hospitals and asylums, an analysis of the air might be made periodically, as the medical superintendent would, in most cases, be qualified to judge if the tests were correctly made; if he did not make them himself. In other buildings, such as churches, halls, schools, workrooms, &c., there would be some difficulty in securing reliable analysis, there being so many conditions under which tests may be made that might render them misleading, and valueless as indicating the general purity of the air.

"It is notorious that, in spite of elaborate tables or analysis of the air of many public buildings in which the most complete forms of artificial ventilation are in use, the condition of the air in these buildings is not found by those occupying them to be in accord with the analysis, so far as their sensations are concerned.

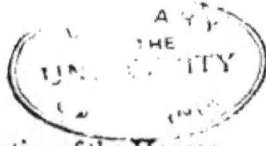

"In a recent report on the ventilation of the Houses of Parliament the following passage occurs: 'When one considers the enormous volume of air, equal to ten times the cubic contents of the House of Commons, which is passed through it every hour, also that the elaborate tables of air analysis are apparently all that could be desired, and yet the ventilation is so notoriously bad, it is clear that there is something very radically wrong with mechanical ventilation by impulsion, and that tables of analysis and of volumes of air passed through a building are not to be accepted as correctly indicating either the general purity of the air, or the efficiency of the ventilation.'

"Analysis of the air is a very good thing so far as it goes, but it may, with some justice, be objected that it does not go far enough, for all the tables of analysis in the world will not convince people that the air of a building is pure if they, from their own actual experience and the unerring test of their senses, are satisfied that it is not so.

"After all, the feelings of the occupants are the surest index of the state of the air in a building and should in most cases be accepted as conclusive, irrespective of air analysis which, however desirable, is not always to be relied upon.

"If the air but smells fresh and sweet, and there is an absence of that feeling of *malaise* and discomfort usually experienced in a badly ventilated building, most people are content to look upon air analysis as a negligible quantity, though they may not always be correct in doing so."

Ventilation Tests. .

"*Reports of ventilation tests are misleading unless
the exact details of the systems tested are given, and the
conditions under which each has been applied and tested
are clearly stated, as one system may have been applied
and tested under the most favourable conditions, whilst
another may have been incompletely applied and tested
under unfavourable conditions. Where these particulars
are not given all such tests must be considered as valueless
as a means of comparison.*"—VENTILATION.

"It is a curious fact, and one that should be
carefully noted, that not one single instance is given
of the mechanical system having been tested against
any automatic system, the schools mechanically
ventilated being tested against schools having *only the
windows and doors to rely upon for ventilation*—that is
to say, schools in which *no system of ventilation existed
at all.* It would have been interesting to have known
what the results would have been had a good automatic system been in use, applied under equally
skilled advice as the mechanical. They (the experiments) are, for all scientific or practical purposes,
valueless, as they prove no more than that an elaborate
and costly system of mechanical ventilation has been
found in a few special instances to be slightly more
effective than where there was *no system of ventilation
in use at all.*"—*Building News.*

Testing Ventilators.

" There is only one trustworthy way of proving the efficiency of any ventilating arrangements, and that is by actual and extended practical experience. Lecture-room experiments, as all privately conducted experiments can only be called, may be all very well in theory, and show certain results, but in actual practice the results are generally found to be very different." — ARCHITECT.

Experiments have been instituted from time to time for the purpose of ascertaining, if possible, the exact relative values of exhaust ventilators, by testing them on specially arranged pipes.

The results of these tests were found, however, to be so conflicting that it is now considered by sanitarians that experiments of this kind are valueless as a means of determining which is the most efficient ventilator under all conditions.

The "Kew experiments," inaugurated by the Sanitary Institute, may be cited as a well-known example.

The *Times*, which was selected by the Institute to publish the report of the tests, said : " The method of testing was incorrect, and therefore the tests are valueless. Neither in the case of the cowls nor of the tubes was their true value as extractors ascertained."

This method of testing has also been discredited by so-called "experiments" carried out by certain cowl makers and patentees of rival systems, when it was found in the case of each that the results were exactly reversed.

There is only one reliable way of testing ventilators, and that is when they are in practical use.

Their efficiency can only be determined when they have been tried for a lengthened period, in every season of the year, and under all conditions of the weather and of the buildings to which they may be applied.

A Natural System of Ventilation applied to a Council Chamber.

"We can quite endorse the favourable opinion of it which has been given by Sir John Monckton, the Town Clerk."—BUILDER.

"Any ordinary observer who has had the privilege of entering the Council Chamber of the Guildhall, London, even when devoid of occupants, would be almost sure to ask himself the question as to how the room was ventilated. It is one of those apartments that appears to strike the mind that a something more than can be seen is necessary for the healthful comfort of those whose civic duties compel them to pass a portion of their time in it, and if the stranger has had an opportunity of being present when the City parliament have been holding their debates, he would have soon found that the atmosphere was anything but inviting; and had he asked any of the officials, or even the members of the Council themselves, he would have heard dissatisfaction expressed on all hands, for it is notorious that, in spite of different attempts to remedy it, including what we may term an elaborate system introduced no later than three years since, the room in which the City Fathers hold their conclaves has been one of the worst ventilated apartments, for its size, in the metropolis. We say 'has been,' for a new system of ventilation has recently been completed that promises to surpass all previous attempts to introduce fresh air and exhaust the vitiated, and one that after many severe tests has received the approbation of all concerned in the carrying out of the plan. Smarting under the annoyances they had long laboured under, the City Architect, by the direction of the Corporation of London, requested Messrs. Robert Boyle & Son, of Holborn Viaduct and Glasgow, whose success as ventilating engineers has become a matter of world-wide notoriety, to submit to him a system of ventilation adapted to the requirements of the Chamber, but on the condition that it would only be accepted after exhaustive trials had proved to the City Architect and a committee that it was successful; and if not to their satisfaction, every vestige of it was to be removed within a specified time, and everything made good at Messrs. Boyle's expense.

"Thus the contract was entered upon on the principle of 'no cure, no pay.' Confident in their system, the firm accepted the terms, and the guarantee, which was of a very stringent character, was drawn up by the Corporation, and was of such a nature as would put Messrs. Boyle's system to a most severe and crucial test. In January last the Chamber was handed over to them to commence operations, and as we have carefully watched the progress of the work, and were present only a few days since—it now having passed through the ordeals prepared for it—we have much pleasure in adding our testimony to its satisfactory results. It has received the approbation of the City Architect, the committee, and, we may add, of the public. Messrs. Boyle have had the privilege accorded them of inviting as many as they pleased to witness the results, and a large number of invitations were issued by them, in which they included those gentlemen they considered interested in the question of ventilation, and competent to pass an opinion upon the work, and also competitors in their own profession, advocates of totally different systems.

"Referring to the extraction of the vitiated air, it is not necessary to give the full table of results from January until the present time; we need only say that the average quantity withdrawn amounted to 500,000 cubic feet per hour, and that during the whole of the experiments, official and otherwise, not the slightest downdraught was experienced: had it been otherwise the conditions would not have been fulfilled, and the firm would have been called upon to remove their appliances, and their system would have been pronounced a failure, one of the principal conditions being that downdraughts should be entirely absent, and that a continuous and powerful updraught should be maintained.

"Previous to Messrs. Boyle undertaking this work great complaints had been made with respect to the draught always present in the Chamber, but since the completion of their arrangements lighted candles have been placed in various parts of the room, the flame has been watched, and it has been clearly demonstrated that they were in all instances perfectly steady, showing that the draughts had ceased to exist.

"After six months' practical experience, Mr. Horace Jones, City Architect and President of the Royal Institute of British Architects, has certified that all the conditions have been fulfilled, and that the ventilation is successful; and, as we before observed, this has been endorsed by the committee, and Messrs. Boyle have received their account. Thus another public building has been added to the long list of those that the firm have successfully ventilated.

"The following report has been received by Messrs. Boyle from Sir John Monckton, the Town Clerk :—

"GUILDHALL, E.C.

"I am asked by Messrs. Boyle & Son to state my personal experience of the recent ventilation of the Council Chamber. I can do so in a very few words. Until the present year I never knew what it was to leave the Guildhall on a 'Common Council day' without a headache. I now do not know what a headache is. In fact the palpable alteration for the better in the atmosphere, even on crowded days, is very satisfactory, and it appears to me—a non-expert—that Messrs. Boyle's system is eminently valuable and practicable.

"In addition, the officials of the Court and many members of the Council have personally thanked Messrs. Boyle for the benefits they have derived, from a health point of view, since the application of the system. From a large number of equally favourable letters received by the firm from independent sources, we abstract the following received from an eminent London physician and sanitarian :— 'After thoroughly testing and examining your system of ventilation as applied to the Council Chamber of the Guildhall, I have great pleasure in being able to say that I am in every way satisfied with the result. It is, in my estimation, the most perfect system yet introduced, and I greatly doubt there being any room for improvement upon it. The fact of your appliances being entirely self-acting and having no movable parts liable to get out of order after having been placed, is a great boon, as it renders any after cost or attention unnecessary. During the time that I have devoted to testing your appliances I have always found them entirely free from downdraught, a feature I have not found in any other system, and which I opine to be of the greatest advantage, as it not only serves to ventilate the building more thoroughly, but furthermore maintains an even temperature therein, which no sudden change of weather can affect. My pursuits have rendered me practically acquainted with many systems of ventilation now in use, both automatic and otherwise, and after careful examination I have arrived at the conclusion that your system is not only the simplest but the most efficacious I have yet met with. This opinion is, in so far as the Council Chamber is concerned, also shared by some friends of mine who are members of the Council, and who pronounce the application of your process simply faultless.'"—*Architect.*

"We have had an opportunity of being present in the Chamber during a meeting of the Common Council, and being aware, from personal experience, of the bad state of the atmosphere which used to prevail on Court days, we are in a position to say that a marked improvement has been effected—so decided a change for the better, in fact, that we can quite endorse the favourable opinion of it which has been given by Sir John Monckton, the Town Clerk."—*Builder.*

" We have witnessed the operation of the system, and can now speak from personal experience of the merits of the principle which Messrs. Boyle have applied to this and other public buildings in London and the provinces. . . . Messrs. Boyle & Son's system is extremely simple, and may be described to be the application of means by which the natural laws of ventilation can be effectively brought into operation. . . . . Our examination of the updraught in the shafts has shown the efficiency of the ventilators."— *Building News.*

Electric Fan Extraction applied to a Council Chamber.

" No architects of great repute or position have of late years relied upon mechanical ventilation."— G. H. BIBBY, F.R.I.B.A.

" Apparently mechanical ventilation is not appreciated by the members of the Bristol City Council, where, according to the *Western Daily Press,* ' Mr. Abbot complained that they were being stifled, the ventilating fan not being in operation. The Mayor gave instructions for the matter to be seen to. Ald. Davies now asked leave to move the adjournment of the house until Tuesday next to the GuildhalL

Really that Council Room was unbearable. Mr.
Levy - Langfield begged to second that. Some
confusion resulted. Complaints of the draughts were
made by many members, and Mr. Pembery asked :
Do you object, Mr. Mayor, to our wearing our hats?
The Mayor : Order, order ! Members sitting on that
side of the room then closed the ventilators from
which the cold air was descending. Mr. Pembery :
Will you, Mr. Mayor, give instructions to stop the
ventilating machine? I would rather stifle than be
killed in this way. The draught is terrible. The
Mayor : If you will kindly take your seat I will have
the matter attended to. It was then understood that
the ventilating fan was put at half speed and business
was resumed.' If the choice is really between stifling
and 'draughts,' the ventilation of the Bristol Council
Chamber is a problem still unsolved."—*Building
News* (February 11th, 1898).

Electric Fan Extraction, &c.

(*Continued.*)

———

"*Ideas regarding the best method to secure ventilation have undergone considerable change within recent
years, mechanical arrangements giving place to simpler
and more effectual methods.*"—BUILDING NEWS.

———

"According to the *Bristol Mercury*, there are
serious complaints amongst the Councillors about
draughts in the Council Chamber. The excessive
draught in the Chamber has led to inquiries, which
have resulted in the explanation that the special

33

electrical ventilation apparatus has never had a real chance of justice being done to it. This arises, the *Mercury* states, 'from the fact that it is so scientifically arranged and adjusted for precise conditions, that the moment these conditions are broken the delicate arrangement of the apparatus is dislocated and its effect entirely lost. It is stated that it was designed for a room with closed doors, and the ordinary ventilators in the walls of the hall intended to admit fresh air should be left open. But the practice is for all these ventilators to be closed, as the room is felt to be too cold, and then only too often the doors at the extreme end are thrown open. Immediately these conditions prevail the whole effect of the electric ventilator is subverted, with the result that an insufferable draught is generated.' "—*Builder* (May 14th, 1898)

Electric Fan Extraction.

"*Mechanical ventilating arrangements are generally utter failures.*"—P. GORDON SMITH, F.R.I.B.A. (Architect to the Local Government Board).

"Under the heading of 'Ventilation,' a correspondence has appeared in the *Manchester City News* relative to the discomforts of electric fan ventilation. The following extract is suggestive, and would seem to confirm the complaints which have reached us from time to time with respect to that method of ventilation :

'The mephitic atmosphere of public rooms, committee rooms, club houses, lecture halls, churches, and the like has long been a source of disease. The application of the electric fan has up to now done little more than substitute a thorough draught, the which of all discomforts an Englishman will least tolerate. In the circulars of ventilating engineers you find the statement that their particular fan will renew the air of a room so many times, say six times, per hour. This statement is not true. What is true is that sufficient air is extracted whose volume would fill the room six times per hour, and that, of course, an equal quantity of air is sucked into the room; but to infer that all the vitiated air, or half, or a third of it had been disturbed and drawn off is to come to a false conclusion. I could take you to a large smoke room in town where a fan is at work on one side of the room where it is so draughty no one will sit. At the other side of the room it is suffocatingly hot, with no apparent movement of air. And this ineffective device cost £200 to put in '"—*Building News* (August 12th, 1898).

Natural *v.* Mechanical Ventilation of Schools.

" The evidence certainly tends to show that artificial ventilation has not proved so satisfactory in actual practice as natural."—BUILDING NEWS.

" After several tests in schools ventilated on each system, it was clearly demonstrated that in none of the

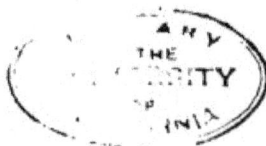

schools examined and ventilated mechanically by extraction—even in a new school opened for a week or two—was the air found to be more pure than in those examined and ventilated naturally without any mechanism.

"Draughts existed in the upper levels of every room ventilated mechanically by extraction, while the halls of such buildings were generally full of draughts."
—*Extract from a paper on " Ventilation" read before the Society of Arts, London* (Feb. 1st, 1893).

"The report on the influenza epidemic presented to Parliament by the Local Government Board indicates the extreme importance of proper ventilation, especially in schools.

"The statistics given point to one town, where the schools are mechanically ventilated on the down-draught principle, as being the 'chief focus' of the disease in Scotland. So far as the children in the schools are concerned this is easily accounted for, as the warm, infected air expelled from the lungs is returned by the descending current, and is not only reinhaled, but is also breathed by the other scholars. This is how infection is spread.

" It is well known to sanitarians that such a mode of changing the air is inimical to health, being not only a direct cause, but a fruitful means of disseminating disease."— *Local Government Journal* on Report to Parliament by the Local Government Board.

Comparative Cost of Natural and Mechanical Systems of Ventilation as applied to Schools.

"The unanswerable objection to the employment of mechanical ventilation is its enormous and unnecessary expense, as it is now generally admitted that mechanical ventilation is not found to be more efficient than ordinary and less expensive methods." — ALFRED FRAMPTON, F.R.I.B.A.

"In naturally ventilated schools the cost of the ventilation is but a fraction. With mechanically ventilated schools this comparison cannot be made. To fit up a school for 1,000 children with mechanical ventilation would cost from £400 to £700 more than if one of the ordinary methods were employed."— PROFESSOR CARNELLEY.

"Mechanical ventilation has not worked satisfactorily, and it is still a question whether we have gained by its adoption anything equivalent to its expense."—*Report from the Clerk to the Birmingham School Board.*

"The expense of mechanical ventilation is unnecessary, and no architects of great repute or position have of late years relied upon mechanical ventilation."—G. H. BIBBY, F.R.I.B.A.

"Architects and School Boards contemplating the adoption of mechanical ventilation would do well to first satisfy themselves that the outlay would be justified by the results, and whether the same, or better results, might not be secured by simpler and less costly methods; the results attained in other quarters certainly point in that direction."—*Building News.*

A Natural System of Ventilation applied to a Public Building.

"The experiments seem to have demonstrated the perfect success of Messrs. Boyle's work." —BUILDER.

"The Long Room of the London Custom House —one of the largest rooms in the world—has recently been ventilated by Messrs. Robert Boyle & Son, Ventilating Engineers, of London and Glasgow, under the direction of Her Majesty's Office of Works. The system applied by them is a combination of their well-known 'Air-Pump' ventilators and vertical tube air inlets. As the Long Room of the Custom House has long been notorious for being one of the worst ventilated rooms in London, and was so officially reported to be by Professor Faraday and Dr. Ure so far back as thirty years ago, considerable interest has been displayed amongst those skilled in sanitary science as to the result of Messrs. Boyle's efforts to remedy the evil. On the completion of the work, a series of experiments were instituted to test the efficiency of the system. A large number of scientists and others attended these experiments. During the tests, the 'Air-Pump' ventilators were found to be extracting an average of 400,000 cubic feet of air per hour.

"It is interesting to notice that, when a dense fog prevailed outside, the atmosphere of the Long Room remained quite clear, and continued so throughout the day. During the whole of the experiments there was not the slightest downdraught discernible in the 'Air-Pump' ventilators, even when all the inlets were closed. All those present at the experiments expressed themselves highly satisfied with the results obtained. The 'Air-Pump' ventilators have in this instance been put to a series of the most severe and searching tests by men the best qualified for such work in this country, and they have undoubtedly maintained, in the most satisfactory and conclusive manner, the high reputation they have long enjoyed as the most efficient foul-air extractors in existence. They have also been tested under conditions more than usually unfavourable for their action, viz., in the middle of winter, with a frosty external atmosphere, and at a time when most so-called self-acting ventilators are usually closed up, to prevent the downdraught they would otherwise admit, as was notoriously the case in this very room under the old ventilating *régime*. With regard to the improvement which the introduction of Messrs. Boyle's system has effected in the atmosphere of the room, those occupying it testify to in the most marked manner—one gentleman, who has been for over thirty-four years engaged in the room, stating that, compared with the past, the present condition of the room was as Heaven to the nether regions. Before Messrs. Boyle's system was applied there used to be a perpetual haze or cloud hanging below the ceiling. This has since entirely disappeared. In foretime, when a fog got into the room, it was several days before it found its way out, even though windows and doors were kept open On entering the room on Monday morning after the dense fog of Sunday, the air was found to be perfectly clear, not a trace of fog being discernible A number of clerks engaged in the room informed us that, under the old arrangement, before the day was half over, they suffered from severe headaches and general prostration caused by the vitiated atmosphere, but since the application of Messrs. Boyle's system they felt almost as fresh when they left in the evening as when they came in the morning, and that they do not now suffer from headaches is certainly a most convincing proof of the efficiency of the ventilation. Sea captains, we are informed, visiting the room, used to complain loudly of the foul state of the atmosphere in which they transacted their business, and now we are given to understand they speak as strongly in favour of the improvement which has been effected. Messrs. Robert Boyle & Son are to be congratulated upon the success they have achieved, and the London Custom House may proudly be added to the long list of public buildings successfully ventilated by them, and which bear testimony to the soundness of the principle they advocate. As founders of the profession of ventilating engineers, they have raised the subject to the dignity of a science, and brought their system to such a state of perfection that we understand they are prepared to guarantee the efficient ventilation of any building entrusted into their hands, no matter how many other systems may have been previously tried and found unsuccessful."— *Industry.*

Report of the Long Room Ventilation Committee.

"I have to inform you that I have conferred with the other members of the Long Room Ventilation Committee, and to state that we are unanimously of the opinion that your ventilation appliances have been a success. It is evident that no system of ventilation in so large a space as the Long Room can be rendered so perfect as to suit all idiosyncrasies and temperaments, but I confidently assert that we have enjoyed, since your appliances have been in action, a purity and clearness of atmosphere to which we had long been strangers."—H. HANCOCK HOOPER, Chairman, Long Room Ventilation Committee, H. M. Customs, London.

"Messrs. Boyle are to be congratulated upon having successfully grappled with a grave difficulty, which had almost become a public scandal."—*Civilian.* the accredited organ of the Civil Service.

Ventilation by Propulsion
(" Plenum " System),
as applied to Public Buildings.

" I never yet knew of a system of propulsion, pure and simple, that effected an efficient and satisfactory ventilation of any large building."—REPORT ON THE VENTILATION OF THE HOUSES OF PARLIAMENT.

"Notorious examples of the failure of mechanical ventilation by propulsion, or artificially forcing air into a building, are the new London Law Courts and the Houses of Parliament, where mechanical arrangements are employed ; and all that money and the highest scientific and engineering skill could do to make the ventilation satisfactory has been tried, but without success.

"Mechanical ventilation by propulsion forces air into a building under pressure and at a high velocity—destructive of diffusion—causing disagreeable and dangerous draughts in the vicinity, and in the line of the inlets and outlets, the other parts of the building being left wholly unventilated, as the incoming columns of air usually travel, or are propelled, in a direct line to the nearest outlet, and there make their escape. Engine-driving columns of air through a building is not ventilating it. It is a well-known fact that those employed in public buildings mechanically ventilated by propulsion constantly complain of feeling ill, and are frequently incapacitated from fulfilling their duties."—*Building News.*

A Natural System of Ventilation applied to a Hospital.

"*For Hospitals natural ventilation certainly seems the proper plan.*"—PARKES.

"*The system has the great merit of being automatic.*" —BUILDER.

"One of the wisest acts of the Metropolitan Asylums Board since its formation was the acquirement of the *Castalia* (which was originally built on the twin principle for the prevention of sea sickness), and the conversion of the vessel into a small-pox hospital on the Thames. Knowing that in many hospitals there is inefficient ventilation, the Metropolitan Asylums Board invoked the aid of several experts on the subject, Professor de Chaumont being the principal adviser. After careful inquiry, it was decided to adopt Messrs. Robert Boyle & Son's system of ventilation, and the Local Government Board approved of the selection. This is considered to be one of the most unique examples of ventilation in this country, or indeed in the world. For the extraction of the vitiated air there are provided twenty of the 'Air-Pump' ventilators. Fresh air is admitted all round the wards by means of openings cut through the walls at the floor level. The air passes over hot-water pipes which are enclosed in a false skirting made of iron, perforated at the top to permit of the air being equally and imperceptibly filtered in and diffused throughout the wards.

"Several scientific and medical men watched the progress of the work, and much interest was excited as to how it would answer. Experiments were therefore instituted by the Board in order to test the efficiency of the system. The results were most satisfactory ; indeed, they are stated to have been far beyond anything that was anticipated. After an extended series of experiments to test the 'Air-Pump' ventilators under atmospheric conditions, such as when there was a good wind blowing, and when there was no wind at all, it was found that the ventilators extracted at the average rate of 5,000,000 (five million) cubic feet of air per hour, the air in the wards being entirely changed once every five minutes, whilst there was not the least disagreeable draught.

During the whole of the tests no appearance of a downdraught was found in the ventilators. Several anemometers were placed in the shafts of the ventilators, and the readings were taken every two hours. Anemometers were also fixed outside to register the velocity of the wind. Messrs. Boyle were not present at any of these tests except the first, the engineers and experts appointed by the Asylums Board being alone entrusted with the trials. Dr. Bridges, H.M. Chief Inspector of Hospitals, after carefully examining and testing the system, expressed his entire approval of its action, informing Messrs. Boyle that even when tested in a calm, he found a considerable updraught in the shafts, and at no time any downdraught.

"From their practical character, the value of these experiments must be very great, as they demonstrate the true worth of Messrs. Boyle's system and its capabilities."—*Architect.*

Ventilation of Hospitals by Mechanical Propulsion.

("Plenum" System.)

———

"Any method of ventilation which depends upon mechanical or artificial means for its action cannot be reliable, and therefore is not to be recommended." —PROFESSOR CORFIELD (Professor of Hygiene and Public Health, University College, London).

———

"Experience would seem to justify the hesitation which has been felt with respect to artificial ventilation.

The following quotation from the Report of the Barracks and Hospital Improvement Commission explains this partly :—

" ' In one hospital we examined, which was ventilated by one of the most perfect apparatus [mechanical] we have anywhere seen, and which professed to supply between 4,000 and 5,000 cubic feet of air per bed per hour, we found the atmosphere of the wards stagnant and foul to a degree we have hardly ever met with elsewhere. We at once pointed out this circumstance. An inquiry was immediately instituted, when it appeared that one of the valves of the supply pipe had been tampered with, for no other reason, that we could perceive, except to save fuel by diminishing the quantity of warm air supplied to the sick. The ventilation, in this case, was worse than a delusion.'

" The writer has visited, on several different occasions, three of the important hospitals in Europe and the United States of America in which the ventilation depended on propulsion, and on every occasion the propulsion happened to be out of use for the time.

" Methods of artificial ventilation are more or less dependent upon careful training in the assistants. They may answer well when first put into operation, but the arrangements, in their simplest form, present some complications and require some special knowledge for their efficient working. Hence the changes in *personnel* which necessarily take place in the course of time may introduce want of appreciation or of care in the management. Moreover, the continuous cost of working presses upon the resources of voluntary hospitals.

"The more the question is examined, the more advisable does it appear to adhere to simplicity in all details of hospital construction. . . .

"The author visited a hospital recently in which the ventilation was by propulsion. The amount of fresh air which was entering the wards was stated to be at the time at a rate of over 5,000 cubic feet per patient per hour, and yet there was a distinct feeling of relief and freshness on passing from the ward to the open air. . . .

"Surgeon-General Billings of the United States Army, mentioned an experiment in the Barnes Hospital, Washington, where fresh air inlets for warmed air were placed near the ceiling, and extraction outlets in the floor.

"In this experiment it was found, that when warm air was admitted near the ceiling there was a difference of ten degrees in the temperature between the floor and the ceiling, and that the patients complained of cold feet and discomfort.

"Surgeon-General Billings also remarks that when the warm air is introduced near the ceiling it is impossible to vary the temperature at different beds, a thing which it is often desirable to accomplish in a hospital."—*Hospital Construction*, 1893 (SIR DOUGLAS GALTON, F.R.S.).

Mechanical Ventilation by Propulsion with Extraction.

"There would be a difficulty in effectively ventilating hospitals by propulsion. The results might easily be mischievous rather than beneficial."—DR. FRANCIS VACHER.

York County Hospital.—"An instructive history belongs to this hospital, which is one of the most interesting examples of the complete failure of artificial ventilation in this country. The air, after being warmed over a series of hot pipes, was driven into the wards by mechanical means, and the foul air sucked out by some form of aspirator. The wards were always close and sickly, and even offensive. The patients complained of the deprivation of fresh air, and the medical officers also complained of the state of the hospital. Worse than this, the health of the patients suffered, especially those who were submitted to any operation, however trifling. In a word, the hospital went from bad to worse, until the surgeons abandoned operations of all kinds rather than incur the almost certain risk of a fatal termination. So at last the hospital had to be emptied and cleansed, and the elaborate apparatus abandoned. Since that time erysipelas as a hospital disease has disappeared and operations do well."

Bristol General Hospital.—"At this hospital an apparatus was set up which consisted of a shaft in the garden from whence air was drawn into the basement, where it was heated by passing over a series of hot pipes ; from thence the warm air went to the wards. The foul air from the wards was drawn out by a series of shafts communicating with a central tower in which an up-current was induced by heated flues, but after a short trial it was found that the hospital was becoming infected with erysipelas, and it was abandoned."—*Hospitals and Asylums of the World,* 1893 (SIR HENRY BURDETT).

Extracts from a Paper on Ventilation read by

Mr. Keith D. Young, F.R.I.B.A.,

before the Architectural Association, London,

November 23rd, 1894.

"*The system of propulsion for hospital ventilation has not found general favour with hospital architects or managers in this country.*"— SIR DOUGLAS GALTON, F.R.S.

"The question which concerns us is this: Is there any evidence to prove that wards cannot be kept sufficiently sweet and healthy without recourse to expensive appliances? So far as the experience of hospitals in this country goes, I think that not only is there no such evidence, but that there is some very definite evidence in support of the exact converse.

"There is a case cited in the report to which I have before alluded, which goes to prove that mechanical ventilation can become a positive evil . . . a system, moreover, which must involve the construction of a number of shafts, all of which, unless kept scrupulously clean, must become harbours for filth, and a constant menace to the health of the patients.

"A distinguished Paris surgeon, Dr. Le Fort, published a report on Hospital Hygiene. He there compares the mortality in London hospitals with that of the Paris hospitals, very much to the disadvantage of the latter, and professes himself decidedly in favour of the natural means of ventilation adopted in London to the artificial systems in vogue in some Paris hospitals. But he says one need not go to England to search for means of comparison between the two systems ; for the two hospitals in Paris where the mortality is greatest are precisely those in which artificial ventilation is employed.

" In the discussion which followed,

" Mr. Alexander Graham, F.R.I.B.A., said he was very glad to hear Mr. Young advocate natural means of ventilation. . . . He was pleased to hear Mr. Young condemn some of the mechanical systems, for he never heard of one which was perfect in action, and he was sure they were very expensive.

" Mr. Gordon Smith, F.R.I.B.A. (Architect to the Local Government Board), said in the main he endorsed all that Mr. Young had said in his condemnation of mechanical ventilation.

" Dr. E. A. Fardon (Resident Medical Officer,

Middlesex Hospital) held that, unless it were
absolutely impossible to secure a natural system of
ventilation, owing to the unfortunate situation of a
building, there was no excuse for adopting an artificial
method. He had inspected many mechanical systems
of ventilation, including those at the Houses of
Parliament and the Law Courts, and had never seen
one that was approved by those who used it. He
thought the importance attached to uneven tem-
peratures in wards was overrated."

Birmingham General Hospital. "When visiting
(Downdraught "Plenum" System.) this hospital I
looked down
the warm fresh air inlets (of the Children's Surgical
Ward) and there saw a collection of rubbish. . . .
The air certainly lacked freshness."—A. S. E.
ACKERMAN, M.I.C.E.

Extracts from an Article in THE HOSPITAL
(*April 1st,* 1899) *by* ALICE RAVENHILL *on the
Mechanical Ventilation of Birmingham General
Hospital.*

Birmingham General Hospital. "One of the
(Downdraught "Plenum" System.) largest installations laid down
in England for mechanical ventilation is that on the
'Plenum' system at the new General Hospital at
Birmingham. . . . To ensure the successful
application of the system the windows are hermetically
sealed and serve for lighting purposes only. . . .
The great cost of such installations as this at Birming-

ham, and the natural desire to learn whether results
justify the outlay involved, lead to the working of the
' Plenum' system in large institutions being keenly
watched by both medical and sanitary authorities.
. . . Those sceptical as to its advantages bring
forward instances of its uncertain working—one in-
stallation proving a success, another nominally on
identical lines verging on the reverse. . . . The
visit in question was paid with an open mind, and in
conjunction with others anxious to observe in practice
a previously carefully studied theory. . . . The
point most generally commented upon was the
apparently unnatural stillness of the atmosphere,
whereas the air was being much more rapidly changed
than is usual. . . . It seems that this subjective
sensation is almost invariably experienced by new-
comers to the hospital, though both nurses and
patients gradually become accustomed to what
scientists assert should be a natural (but which to
most of us at present appears to be a somewhat
artificial) atmosphere. It has been stated that this
feeling of oppression does not meet with the entire
approval of the visiting staff, as exercising upon them
an indefinably depressing effect. The atmosphere of
those wards visited struck the party as somewhat close,
lacking in freshness ; . . .

"The kitchens are alone well worth a visit, but
again the first impression experienced was a craving
for the familiar, but forbidden draught. . . . Some
of the party confessed, with shame and confusion of
face, their preference for the old-fashioned system,
which preached the free exposure to the outside air of

all vessels used in connection with food, rather than their bestowal in closed chests or in a confined atmo sphere.

"Where the initial outlay on hospital construction is so compulsorily a matter of supreme importance, it would seem advisable to secure very convincing proofs of the advantages derived from the mechanical ventilation in force at Glasgow and Birmingham, before incurring the cost of installing the ' Plenum ' system in other hospitals. It must be borne in mind that with an allowance of 1,000 cubic feet of air space per bed, adequate ventilation by natural means is practically always possible. . . . The experience gained by visiting several schools where the ' Plenum ' system is installed tends even then to the conviction that complete freshness is only assured when mechanical methods of changing the air can be supplemented by the 'natural' opening of windows and doors during the recreation intervals, permitting thorough flushing of classroom and hall, otherwise a stale, stuffy smell is usually unpleasantly perceptible to sensitive nostrils.

"Depending, as it does, on highly skilled management for its efficiency, some scientific training would seem desirable for its controllers. For instance, if comment be made to those in charge as to the closeness of schoolroom or ward, no such confusion of ideas should exist as to cause the temperature to be quoted in proof of the satisfactory condition of the air. So long as but one person in charge of a part of any building thus ventilated, fails to recognise that the thermometer registers temperature, but does not gauge atmospheric impurities, so long will an element of

possible danger to health exist for those dependent on
this and kindred means of air purification. . . .

"In conclusion, the moral consequence of seeing
only closed windows upon a large body of children or
of ignorant patients seem worth consideration, suscep-
tible as is this objection to being dismissed as
purely sentimental."

Sir Douglas Galton, F.R.S., *on the comparative
efficiency of "Natural" Ventilation and of Mechanical
Ventilation on the "Plenum" system as applied to
Birmingham Hospital and the Victoria Hospital,
Glasgow.*

Victoria Hospital, Glasgow.
(Downdraught "Plenum" System.) "Wherever a
building is occupied per-
manently with definite conditions it is far better to
trust to the ventilation of Nature—natural ventilation
—than to the artificial pumping in of air. During that
part of the year when the windows can be opened, at
least fifteen to twenty times as much fresh air would
be admitted into the ward, per patient, than could be
obtained if the air was pumped in, as in the Birmingham
Hospital, and that was certainly much more beneficial to
the patients. I visited the Victoria Hospital at Glasgow,
and went through the wards. I then went out with
the matron or head nurse, into the Nursing Sisters'
Home, which was attached to the hospital. I said to
the matron, 'Why do you not, if you speak so highly
of the system of forcing the air into these wards as
being so very desirable, adopt it in your own Nursing
Homes?' and she replied, 'We find the open windows
so very much more refreshing.'"

Extract from the Discussion on a paper read before the Royal Institute of British Architects, May 29th, 1899, when the results of an inspection of the " Plenum" system as applied to a Birmingham Hospital were stated.

A Birmingham Hospital.

(Downdraught "Plenum" System.)

Mr. C. W. Mountford, F R I.B A.,

said :—" Only quite recently several members of the Institute Council were taken over a large building in Birmingham which had recently been fitted up with this system, and he had been struck with the considerable accumulation of very black dust in the ducts through which the warmed air was forced in the various wards. Now, that dust represented simply the thicker and heavier particles which the fans had not been able to force into the wards, and it seemed to him an absolute certainty that a great deal of dust had been forced into the wards and swallowed by the patients. In a new building there was not much objection to that, possibly, but as time went on and the soot accumulated, and more and more was forced into the lungs of the patients, it would become a very big question whether the 'Plenum' system would not be taken out of those buildings and something of a different nature put in. It was said of the 'Plenum' system that it was equally available in summer for forcing in cool air. That, he thought, was a great mistake. People did not mind having air forced into a room so long as that air was very warm, but when they began to force in cold air the inevitable result would be that the occupants of the room would block up the openings."

Air Screens:

Their comparative efficiency with Natural and Mechanical Systems of Ventilation.

"If air is forced rapidly through a screen it cannot fail to carry dust with it."—Sir Douglas Galton, F.R.S.

"There are many different contrivances in use for washing, screening, and purifying the air, the majority of which answer their purpose fairly well, though the evidence would seem to indicate that they are found to be most effective when employed in conjunction with a 'natural' system of ventilation where the air is not mechanically propelled, but is drawn through by extraction, the air being more thoroughly purified moving in at a *low* velocity than when forced in under pressure.

". . . . The Metropolitan Asylums Board employ a natural system of ventilation by extraction from the ceiling, and admission of air from below, —the plan approved by the Local Government Board . . . . The purity and freshness of the air in the wards, especially in the more recent hospitals, is remarkable.

". . . The fresh air supply is screened and purified, and the temperature can be regulated as required. No fog or other impurities in the external air ever find access to the wards, the atmosphere of which is perfectly clear and bright even during a dense fog."—*Local Government Journal.*

Victoria Hospital, Glasgow. "A close examination of, and several
(Downdraught "Plenum" System.)
tests which we applied to, the system of ventilation
at the Victoria General Infirmary convince us that,
whereas the main duct is kept comparatively free
from grit and dust, the small ducts communicating
directly with the wards have not been protected
against the intrusion of these deleterious matters.
Indeed, the evidence shows that with the air very
much enters that should be excluded from the wards,
as the walls and ceilings eloquently testify. We are
of the opinion that the method of washing the air
is imperfect, and might be improved. This would
tend to minimise the amount of dust and blacks
which at present find their way into the wards."—
Hospitals and Asylums of the World, 1893 (SIR HENRY
BURDETT).

"Plenum" System. — "Unfortunately I have
lived for the last dozen years or more under a system in
which air was forced in, and anything more pernicious
I cannot imagine, in London, at all events. It might
be satisfactory perhaps in the dustless atmosphere of
the country, but in the dust-laden air of the metropolis
such a system was abominable, and my experience
would lead me to prefer in all cases the introduction
of air by suction rather than by pressure. That
system has the great advantage that the air can be
introduced at a very large number of places without
producing draughts."—Dr. ARMSTRONG (Member of
Royal Commission on Ventilation).

Birmingham Hospital. "Only quite
(Downdraught "Plenum" System.) recently
several members of the Institute Council were taken over the Birmingham General Hospital, which is supposed to be the best example of this system of ventilation, and Mr. Mountford was struck with the considerable accumulation of very black dust in the ducts through which the warmed air is forced into the various wards. As this dust represents simply the thicker and heavier particles which the fans had not been able to force into the wards, it looked as if a good deal of dust must have passed into the building and been swallowed by the patients. Curiously enough the air forced into the hospital is supposed to be automatically washed and filtered by being made to pass through cocoa-nut matting screens kept periodically wetted."— VENTILATION AND WARMING.

Natural *v.* Mechanical Ventilation.

" Mechanical ventilation is not only costly to instal and maintain, but it is unreliable and perpetually going wrong, either through the want of proper attention (which is seldom given), or owing to the mechanism getting out of order and breaking down."—HOUGHTON.

" My experience of the process of forcing air into buildings is not in its favour. The only safe and sound means for the supply of air is the natural one of obtaining it from a pure source in a free and natural flow." — SURGEON-GENERAL SIR THOMAS CRAWFORD.

" Ventilation can only be successfully accom-

plished at all times when it is effected without
assistance from mechanical or artificial contrivances.
However perfect these may appear, they can never
achieve results superior to those insured by judicious
and intelligent adaptation of natural means, and
they necessarily suffer from the very serious disadvantage that they are liable to interruption without
warning, and with possibly disastrous consequences."
—PROFESSOR WADE.

Cost of Mechanical and Natural Systems of Ventilation as applied to Hospitals.

———

" *The expense of mechanical ventilation is unnecessary, for there is sufficient evidence to show that such
buildings as asylums, workhouses, and hospitals are best
ventilated by 'natural' means, and no architects of
great repute or position have of late years relied upon
mechanical ventilation.*"—G. H. BIBBY, F.R.I.B.A.

———

" The really important point to be kept in view in
regard to ventilation is, that before any system
depending upon mechanical contrivances can be
pronounced worthy of adoption, it must be demonstrated beyond dispute that it is not only as good as
ordinary methods, but appreciably better. For
nothing but a substantial improvement would justify
the largely increased cost, both of construction and
maintenance, necessarily consequent on the adoption
of mechanical ventilation."— *Hospitals and Asylums of
the World*, 1893 (SIR HENRY BURDETT).

E

"The question which concerns us is this : Is there any evidence to prove that wards cannot be kept sufficiently sweet and healthy without recourse to expensive appliances? So far as the experience of hospitals in this country goes, I think that not only is there no such evidence, but that there is some very definite evidence in support of the exact converse."— KEITH D. YOUNG, F.R.I.B.A.

"The unanswerable objection to the employment of mechanical ventilation is its enormous and unnecessary expense, as it is now generally admitted that mechanical ventilation is not found to be more efficient than ordinary and less expensive methods."— ALFRED FRAMPTON, F.R.I.B.A.

Hot Air Aspiration.

" This form of ventilation is not in accordance with the views of modern engineers, or with their ordinary practice."—VENTILATION.

Guy's Hospital, London.—"This building is provided with an elaborate system of ventilation. The fresh air is taken in at the top of two towers, and drawn down to the basement, where it is warmed and sent up to the wards, and admitted at the floor level, and the foul air is extracted from the ceiling level and drawn off through shafts to the top of a lofty tower by means of heat. The system has been anything but a success, and in the report of Dr. Bristowe and Mr. Holmes it is said to be condemned unanimously by the medical staff, who find it utterly inefficient, and who regard these wards as the least healthy in the hospital.

" . . . Having regard to the temperate character of the British climate, we have yet to be convinced that it is desirable or necessary to introduce artificial ventilation into our hospitals. . . . At Guy's Hospital (new building) the artificial system can be compared with the natural system in the older wards; and the result of the comparison certainly does not make for the superiority of the former method."—*Hospitals and Asylums of the World*, 1893 (SIR HENRY BURDETT).

" The Western Infirmary, Glasgow, is ventilated by means of hot water coils placed in shafts communicating with Louvred turrets on the roof, in which are placed hot water cylinders; on testing this system a strong downdraught was experienced, flocks of cotton wool being driven with considerable force down the shafts. At no time during the experiments was there any abatement in the downdraught, or the slightest tendency to an updraught."— *Ventilation.*

"St. Mary's Hospital, London, affords some instructive lessons in ventilation. In their report on this hospital, Dr. Bristowe and Mr. Holmes referred to the existence of a central shaft for extracting the foul air from the wards in which a fire was always kept burning. They said: 'The Secretary informed us, however, that there was reason to suppose the whole apparatus is a failure; that Dr. Sanderson had made numerous experiments which led him to the conclusion that the orifices of exit in the wards act little, if at all, and very often admit air instead of carrying it off, and that there is, in fact, rather a circulation of air in the shaft than an escape of air from it.' In summing up their conclusions these gentlemen said : 'The hospital cannot certainly be regarded as a healthy one, for most of the diseases which constitute unhealthiness in a hospital seem to prevail in an unusual degree, and there seems also to be considerable spread of infectious fevers.' "—*Hospitals and Asylums of the World*, 1893 (SIR HENRY BURDETT).

Downward Ventilation by Propulsion.

"*It goes without saying that fresh air passing through a stratum in which gas lights are burning cannot avoid bringing the carbon oxides, acetylene, and other products of combustion into the respiratory atmosphere. For English practice, therefore, it may be regarded as incontrovertible that, save in a few exceptional instances, an upward current of the fresh air is preferable.*"— BUILDER (September 5th, 1896).

Extract from Report laid before the United States Congress by the Government Commission on Ventilation :—

"The relative merits of the upward *versus* the downward systems of ventilation may be estimated from the following considerations :—

" 1. The direction of the currents of air from the human body is, under ordinary conditions, upwards, owing to the heat of the body. This current is an assistance to upward, and an obstacle to downward, ventilation.

" 2. The heat from all gas flames used for lighting tends to assist upward ventilation, but elaborate arrangements must be made to prevent contamination of the air by the lights if the downward method be adopted.

" 3. In large rooms an enormous quantity of air must be introduced in the downward method if the occupants are to breathe pure fresh air, or about three times the amount which is found to give satisfactory results with the upward method.

" 4. In halls arranged with galleries, the difficulty of so arranging downward currents that, on the one hand, the air rendered impure in the galleries shall not contaminate that which is descending to supply the main floor below, and, on the other hand, the supply for the floor shall not be drawn aside to the galleries, is so great that it is almost an impossibility to effect it.

" Perfect ventilation would not be obtained, for this would only provide for the *dilution* of the impure air, while in perfect ventilation the impurities are not so diluted, but completely removed as fast as formed, so that no man can inspire any air which has shortly before been in his own lungs or in those of his neighbour.

" For these and other reasons the Board are of opinion that the upward method should be preferred."

Extract from Report on the Ventilation of the Chamber of Deputies, Paris, by M. EMIL TRÉLAT :—

" The fresh air is delivered at the top of the room, and the currents of air are strong and varied, causing great discomfort.

"The air is delivered by fan propulsion at the ceiling, and is taken out at the floor, giving a downward system of ventilation.

"The apparatus is powerful enough to change the air in the chamber every six minutes; but it can only be worked slowly, owing to the draughts produced, and the air vitiated by respiration is brought back to be re-inhaled."—EMIL TRÉLAT.

"Whilst the air is in the lungs, it acquires so much heat that it becomes specifically lighter than the surrounding air, and rises above our heads. The heated air which passes upwards should pass away. . . . For the ventilation of rooms exits should be provided for the spent air near the ceiling. . . The method of low ventilation (extraction near the floor) should be avoided on various grounds."—ROYAL COMMISSION ON VENTILATION (Blue Book).

"The down system can never supply really pure air to be breathed by the lungs. The exhalations of the human body are, as they issue, so warm that they must perforce immediately rise. Therefore, if the supply of fresh air comes from above, it can only reach the nose and mouth by driving down with it and mixing with these foul exhalations, and there is unquestionably nothing to breathe except this polluted mixture. In order to keep down the percentage of pollution to a non-dangerous degree, under this system arises, therefore, the necessity of admitting for ventilation fresh air in quantities many times greater than that actually used for breathing, and also a correspondingly extravagant expenditure of heat if this supply be artificially warmed. Thus the only ideally perfect ventilation consists in inducing a regular up-current from a level below that of the human head up to the extraction outlets at the ceiling. Under this system the bulk of fresh air required to be admitted is immensely reduced, as is also the expense of warming it to any degree considered desirable."—PROFESSOR R. H. SMITH.

"The evils of downdraught ventilation cannot be exaggerated. Such a system savours little of the nine-teenth century."—ALFRED FRAMPTON, F.R.I.B.A.

"To the plan of abstracting the foul air near the floor there are at least four grave objections : (1) It is opposed to Nature's law of atmospheric pressure, and therefore requires the use of special abstracting power by means of furnaces for its accomplishment ; (2) by drawing down the foul air it causes it to be breathed over again, which recently-breathed air ought never to be ; (3) the fresh air supplied is apt to be forced in over-heated—in fact, burned—and so made unhealthy ; (4) the long, tortuous flues cannot be kept clean, and will, therefore, become lurking-places for dust and germs. The plan is quite unsuitable for hospitals, and should certainly never be used where there is likely to be infection."—DR. JOHN HAYWARD.

"There are also some statements and theories to which we take exception. One of these is the argument that because carbonic acid gas is 52 per cent. heavier than air, it is, therefore, desirable to ventilate by a downward current, in a room, rather than an upward one. The fallacy of this will be

obvious to every one of our readers who smokes, if he
will notice the course of the same after being exhaled.
It is almost impossible in an ordinary room to make
the smoke go down to the floor without a very violent
effort. In all ordinary expulsion it would be noticed
that the smoke ascends, and quickly becomes dis-
seminated through the air of the room. The scientific
fact is that air exhaled from the lungs is, at the
moment of exhalation, equal in weight to pure air at a
temperature of 90 deg. Fahr. And until the exhaled
air has parted with so much of its heat as to become
heavier than pure air at this temperature, it will rise in
a normally pure atmosphere."—*Builder* (April 30th,
1898).

Hot Air Heating.

*" The hygienic drawbacks of hot air heating are
recognised."*—BUILDER.

" The teachings of scientists and sanitarians as
to the healthiest mode of heating and ventilating
buildings are now being recognised and conformed
to by architects and engineers in a manner that
augurs well for the future.

" The latest evidence submitted at the Royal
Institute of British Architects and at the International
Congress of Hygiene, Budapest, has conclusively
demonstrated that heated air employed for the
double purpose of ventilation and warming is not
only unhealthy, but fails as an efficient heating
medium.

" Sir Douglas Galton, in a paper read at the
Congress, strongly condemned this mode of heating
and ventilating a building, stating that: ' The method

of warming the walls by means of heated air necessarily leaves the walls colder than the air of the
room, and the heat of the body is radiated to the
colder walls. Hence, if the walls are to be warmed
by the air admitted to the room, the temperature of
the warmed air must be raised beyond what is either
comfortable or healthy for breathing, and thus, if you
obtain your heat by warmed air alone admitted direct
to the room, discomfort in one form or the other
can with difficulty be avoided.'

"Professor Corfield (Professor of Hygiene and
Public Health, London University), said : ' Heating
should be done by means of radiant heat, and not by
means of air previously warmed. If air was previously
warmed it would lose a portion of its oxygen, and if
we got air short of oxygen we had to breathe a greater
number of times to supply the required amount, and
that meant more effort.'

" M. Emil Trélat, who is recognised as the highest
authority on the subject in France, and whose report
on the failure of the mechanical ventilation at the
Chamber of Deputies, Paris, we have referred to,
declared that ' the solution of the problem of heating
dwellings had absolutely no connection with that of
their ventilation.'

" Mr. Thomas Blashill (Architect to the London
County Council) and others present at the Congress,
expressed themselves in similar terms with regard to
the unhealthiness of breathing warmed air which was
also employed for the purposes of heating.

"It is now being more generally recognised in
this country that the ventilation and the heating of a
building should be kept separate and distinct, and

that the fresh air warming arrangements should be independent of those employed for heating.

"In the not very distant future we may expect to see the systems which combine the two as obsolete and out of date as the deadly bell-trap or the poisonous pan closet, both of which in their day were upheld as infallible; and although the enlightened principles now being inculcated and adopted have received, and will doubtless continue to receive, the most determined opposition from the makers of ventilating and heating arrangements on the hot air plan, these opponents to progress will sooner or later be compelled to conform to the requirements of advanced sanitary science or be left behind, as mere trade interests can never be allowed to subserve and override those of the public health, which are paramount and must always take precedence."—*Local Government Journal.*

"A most instructive historical fact is the present gradual abandonment in the States and Canada of the hot-air system of house-warming, which was for so long popular, in favour of hot-water pipe and other 'radiator' warming. It is, in fact, impossible that the human body should absorb heat from the air in which it is immersed if that air be not made oppressively hot."— PROFESSOR R. H. SMITH.

"When a building is ventilated with hot air, which also constitutes the means of heating, to satisfactorily effect the latter it is necessary to raise the air to such a high temperature that the oxygen is partially destroyed and its life-giving qualities considerably reduced. There is also experienced with this method a feeling of closeness or want of freshness, even though the air is being rapidly changed."— *Building News.*

" The fresh air supply if heated to such a temperature as is necessary to effectually warm a building, is thereby seriously deteriorated for breathing purposes; therefore, to secure healthy ventilation, the heating of a building should always be separate and distinct from that of the air supply."—VENTILATION.

" Health is only possible when to other conditions is added that of a proper supply of pure air. No one who has paid any attention to the conditions of health, and the recovery from disease, can doubt that impurity of the air marvellously affects the first, and influences and sometimes even regulates the second."—PARKES.

SUPPLEMENT.

REPRODUCTIONS OF DIAGRAMS ILLUSTRATING
THE ACTION OF

Natural and Mechanical Methods of Ventilation

WH'CH

APPEARED IN A SERIES OF PAMPHLETS CONTAINING
REPRINTS OF ARTICLES ON

VENTILATION

PUBLISHED IN THE

Local Government Journal

Sept. 1st, Oct. 13th, Dec. 15th, 1894, and Feb. 16th, 1895.

EXTRACTS FROM WHICH ARE HERE GIVEN
BY PERMISSION OF THE PUBLISHERS.

Congress of the Sanitary Institute on Mechanical Ventilation.

"We have a striking instance at the Congress of the Sanitary Institute, of how an assembly of scientific experts, having an intimate practical knowledge of ventilating systems and their respective values, unanimously condemned what they knew to be a dangerous system, but which non-experts, having no, or at most but a superficial, knowledge of ventilating arrangements, impressed by the working of elaborate machinery, accept as efficient.

"A paper on ventilation was read at the Congress at Liverpool by the architect of the new General Hospital, Birmingham, the system described being downward ventilation by mechanical propulsion.

"It seems that the special feature of the system as described by the lecturer, for which novelty was claimed, is a screen of a certain construction for washing the air, and which is kept saturated with water.

"The lecturer stated that with this wet screen 'the *remarkable phenomenon* was observed of the air being *moistened* when too dry, and being *dried* when over moist.' With regard to this statement we cannot but agree with the expert who prepared the report on the ventilation of the Houses of Parliament, and who, referring to this gentleman's ideas on ventilation, said, 'So long as architects hold such views, so long will our public buildings stand every chance of being examples of inefficient and defective ventilation.'

"The feeling of those present at the lecture, with regard to the screen and the system of ventilation explained, may be gathered from the following extract from *The Builder's* report of the proceedings.

"'A discussion followed, in which Dr. Vacher, Sir T. Crawford, Surgeon-Major Black, and several other experts took part.

"'Dr. Francis Vacher (Birkenhead) thought there would be a difficulty in effectively ventilating hospitals by propulsion, or, if it were done, that the results might easily be mischievous rather than beneficial. He had recommended moistened screens in a report he had drawn up many years ago.

"'Surgeon-General Sir Thomas Crawford said the scheme proposed was one that required very careful consideration. He had had a very large experience of hospitals, and of screens for purifying the air of buildings. In hot climates our army was served out with apparatus for the circulation of air, and with screens very much like those here proposed. His experience of the process of forcing air into buildings was not in its favour. The difficulty was to regulate the amount of moisture, and he was not sure that they did not add more deleterious matter.

"'The only safe and sound means for the supply of air was the natural one of obtaining it from a pure source in a free and natural flow.'

"The system was condemned as 'a retrograde step, which the Sanitary Institute ought not to endorse; that all mechanical arrangements of the kind were apt to get neglected and fall into disuse, and that no figures had been given to show the actual economy of the scheme proposed.'

"The lecturer replied that all that had been advanced in condemnation of this system 'he had heard hundreds of times.' The decision at the Congress of the Sanitary Institute against mechanical and in favour of natural ventilation, confirms the opinions approved at the International Congress of Hygiene, Budapest, and at the Congress of the British Institute of Public Health.

Evils of Downdraught ·Ventilation.

"The breath from a fever patient being of a higher temperature than that of a person in ordinary health, ascends to a greater height, before being pressed down by the descending current, and is inhaled by those who may be close to the bedside, owing to the spreading and consequent diffusion that takes place. Where the current is upwards, as in natural ventilation, the breath from the patient ascends in an unbroken column, and is immediately drawn away without coming into contact with anyone.

"With a downdraught system the patient must inevitably rebreathe a considerable portion of not only his own exhausted air, but also a certain proportion of that of the patients on each side of him, and of persons standing about his bed.

"The watery vapour of his breath, in which the disease germs are suspended, is likewise precipitated on to the bedclothing by the downward current, and there accumulating, proves not only a means of considerably retarding his recovery,—if it is not indeed the actual cause of his death,—but is a dangerous source of infection to others.

"No amount of air forced on this plan into a building could make the ventilation either efficient or safe.

"MECHANICAL VENTILATION ON THE DOWNDRAUGHT PRINCIPLE, BY IMPULSION, OR THE 'PLENUM' SYSTEM, APPLIED TO A HOSPITAL WARD.

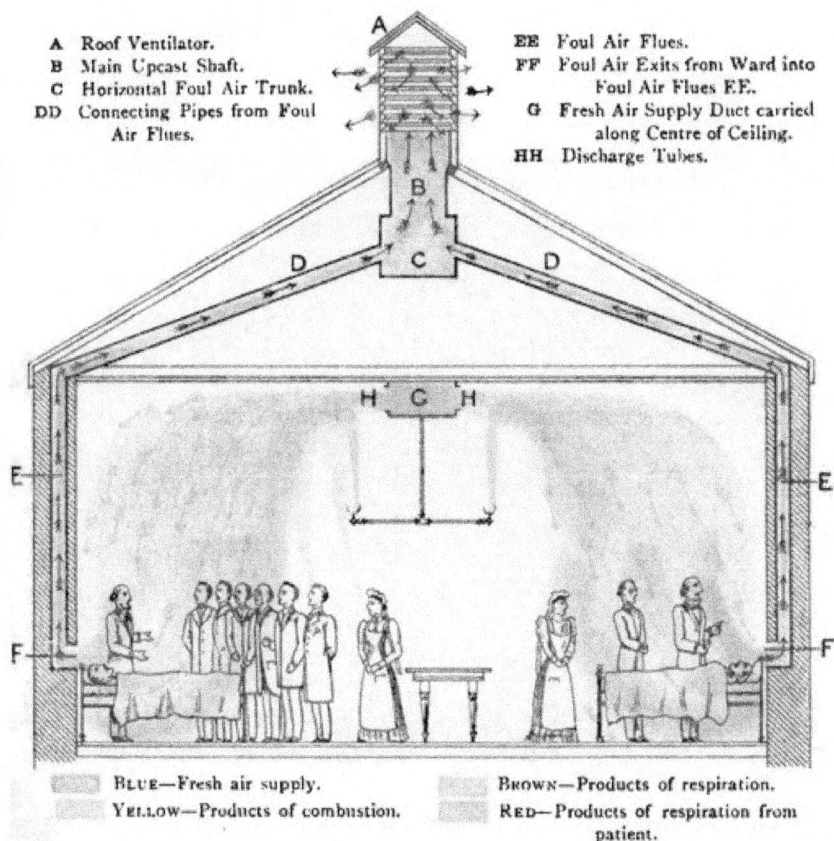

A   Roof Ventilator.
B   Main Upcast Shaft.
C   Horizontal Foul Air Trunk.
DD Connecting Pipes from Foul Air Flues.

EE Foul Air Flues.
FF Foul Air Exits from Ward into Foul Air Flues EE.
G   Fresh Air Supply Duct carried along Centre of Ceiling.
HH Discharge Tubes.

BLUE—Fresh air supply.
YELLOW—Products of combustion.

BROWN—Products of respiration.
RED—Products of respiration from patient.

"Downward ventilation returns the vitiated respired air to be rebreathed, also the poisonous products of combustion. Only an unendurable and highly dangerous down draught could prevent the hot respired air ascending to a height sufficient to ensure its rebreathal. With this system the patients in a hospital inhale foul air only. If the foul air exits are placed either above, below, or at the side of the bed, the results are the same, also when the fresh air is forced in at the upper parts of the side or end walls, or is projected up on to the ceiling."

Comparative Advantages of Upward and Downward Methods of Ventilation.

"In a Government report on ventilation laid before the United States Congress, Clause 3 of the report states :—'In large rooms an enormous quantity of air must be introduced in the downward method if the occupants are to breathe pure, fresh air, or about *three times the amount* which is found to give satisfactory results with the upward method.

"'For these and other reasons the Board are of opinion that the upward method should be preferred.'

"A well-known authority says, 'The evils of downdraught ventilation cannot be exaggerated.' In a published letter to the expert who reported on the ventilation of the Houses of Parliament an eminent architect and scientist says, 'I have before now tried (very much in vain) to drive the idea into people's heads that one can pass an ample supply of fresh air through an apartment without ventilating it.'

"MECHANICAL VENTILATION ON THE DOWNDRAUGHT PRINCIPLE, BY IMPULSION, OR THE 'PLENUM' SYSTEM, APPLIED TO A HALL OR CHURCH.

A Roof Ventilator.
B Main Upcast Shaft.
C Horizontal Foul Air Trunk
DD Connecting Pipes from Foul Air Flues.
EE Foul Air Flues.
FF Foul Air Exits from Church (or Hall) into Foul Air Flues EE.

G Fresh Air Supply Flue.
NOTE.—Several of these Flues are usually constructed in each of the side walls.
HH Inlet Tubes through which the fresh air is mechanically projected at a high velocity on to the ceiling and then descends.

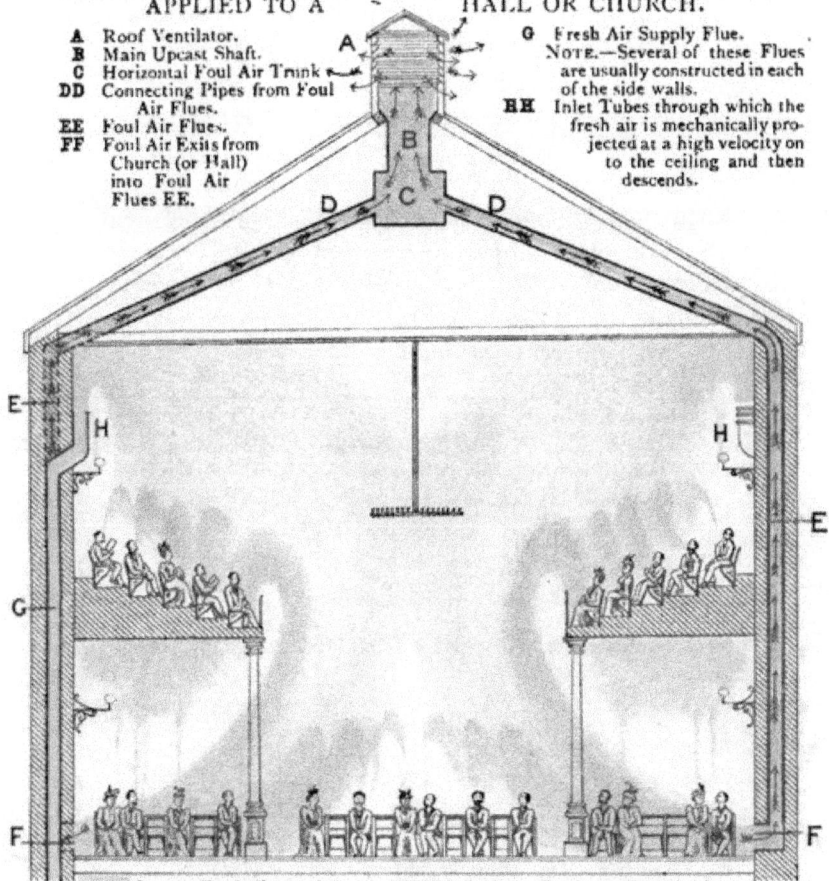

BLUE—Fresh air supply.
YELLOW—Products of combustion.
BROWN—Products of respiration.
RED—Products of respiration from infected persons.

" In halls arranged with galleries, the difficulty of so arranging downward currents that, on the one hand, the air rendered impure in the galleries shall not contaminate that which is descending to supply the main floor below, and, on the other hand, the supply for the floor shall not be drawn aside to the galleries, is so great that it is almost an impossibility to effect it."— Extract from Government Report to Congress on the failure of mechanical ventilation at the Capitol, Washington, U.S.A.

" The air is delivered by fan propulsion and is taken out at the floor, giving a downward system of ventilation. The air vitiated by respiration is brought back to be reinhaled. . . . M. Trélat emphatically condemns downward ventilation."—Extract from Official Report on the failure of mechanical ventilation at the Chamber of Deputies, Paris."

Downdraught Ventilation by Propulsion.

"'This is one of the very earliest forms of mechanical ventilation, devised at a time when the law of diffusion was but imperfectly understood in connection with its relation to ventilation, and when it was the popular belief that carbonic acid gas and other impurities evolved in an occupied building fell to the lower level, and there accumulated in stratas—hence the supposed necessity for expelling the foul air at the floor level, a method now obsolete and out of date.

"   .   Downdraught ventilation by mechanical impulsion is pronounced by sanitarians to be highly prejudicial to health."—*Local Government Journal.*

Natural Ventilation of Schools.

"At the Congress of the British Institute of Public Health, Mr. J. T. Bailey, Architect to the London School Board, read a paper in which he stated his experience of the ventilation of schools, the method which he found most satisfactory being the 'natural' system. Mr. Bailey cited, as an example, the Hugh Myddelton School, opened by H.R.H. the Prince of Wales in December, 1893, which is, as all the London Board schools are, ventilated with a 'natural' system, with which method Mr. Bailey said 'there was an entire absence of what is known as the "school smell,"' and 'these means have proved amply sufficient in their results.'

"MECHANICAL VENTILATION ON THE LATERAL AND DOWNWARD PRINCIPLE, BY IMPULSION, OR THE 'PLENUM' SYSTEM, APPLIED TO A SCHOOL.

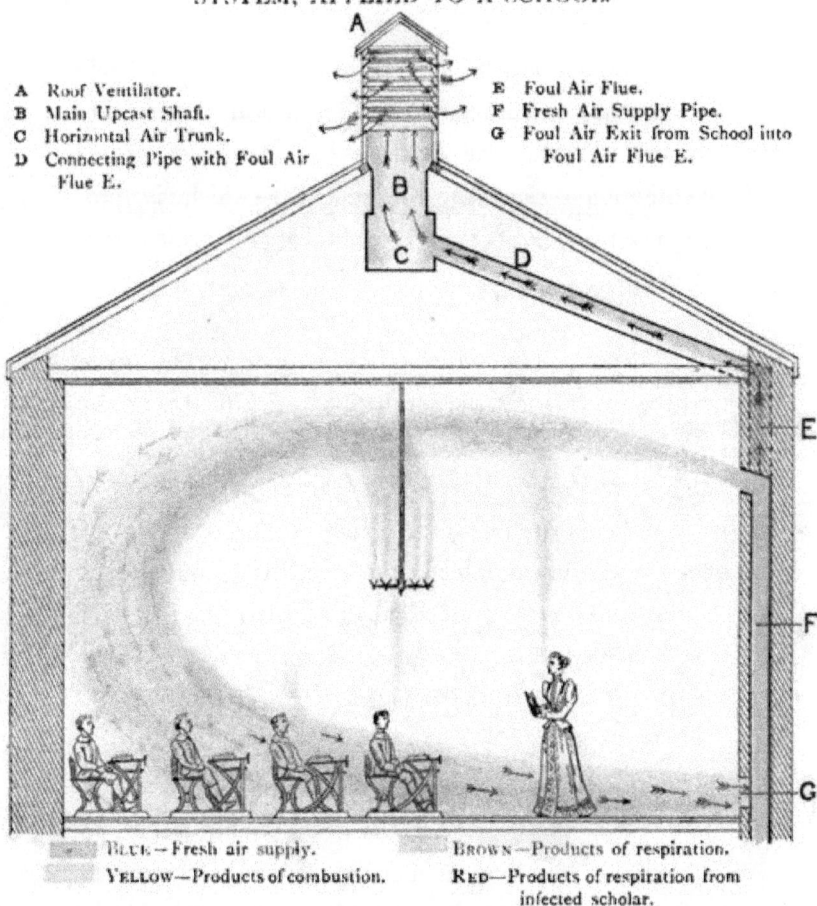

A Roof Ventilator.
B Main Upcast Shaft.
C Horizontal Air Trunk.
D Connecting Pipe with Foul Air Flue E.

E Foul Air Flue.
F Fresh Air Supply Pipe.
G Foul Air Exit from School into Foul Air Flue E.

BLUE—Fresh air supply.
YELLOW—Products of combustion.
BROWN—Products of respiration.
RED—Products of respiration from infected scholar.

"It is well known to sanitarians that such a mode of changing the air (downward ventilation) is inimical to health, being not only a direct cause, but a fruitful means of dissemination of disease, as evidenced by the report issued by the Local Government Board, one town, where the schools are mechanically ventilated on the down draught principle, being specially mentioned as the 'chief focus' of the disease (influenza) in Scotland."—*Local Government Journal* on Report to Parliament by the Local Government Board."

x.

How Infection is Spread.

"The report on the influenza epidemic presented to Parliament by the Local Government Board indicates the extreme importance of proper ventilation — especially in schools — which is pronounced to be the only real safeguard against that disease.

"The statistics given point to one town, where the schools are mechanically ventilated on the downdraught principle, as being the 'chief focus' of the disease in Scotland. So far as the children in the schools are concerned this is easily accounted for, as the warm, infected air expelled from the lungs is returned by the descending current, and is not only reinhaled, but is also breathed by the other scholars. This is how infection is spread.

"With natural ventilation by diffusion and extraction the evils described cannot occur, as the air is admitted at a level, and in a manner in accordance with the laws of Nature, the hot vitiated air being drawn off at the top of the building, to where it ascends, and not permitted to return, as with downward ventilation, to spread disease and death amongst those compelled to breathe it.

"MECHANICAL VENTILATION BY IMPULSION WITH EXTRACTION, APPLIED TO SCHOOLROOMS.

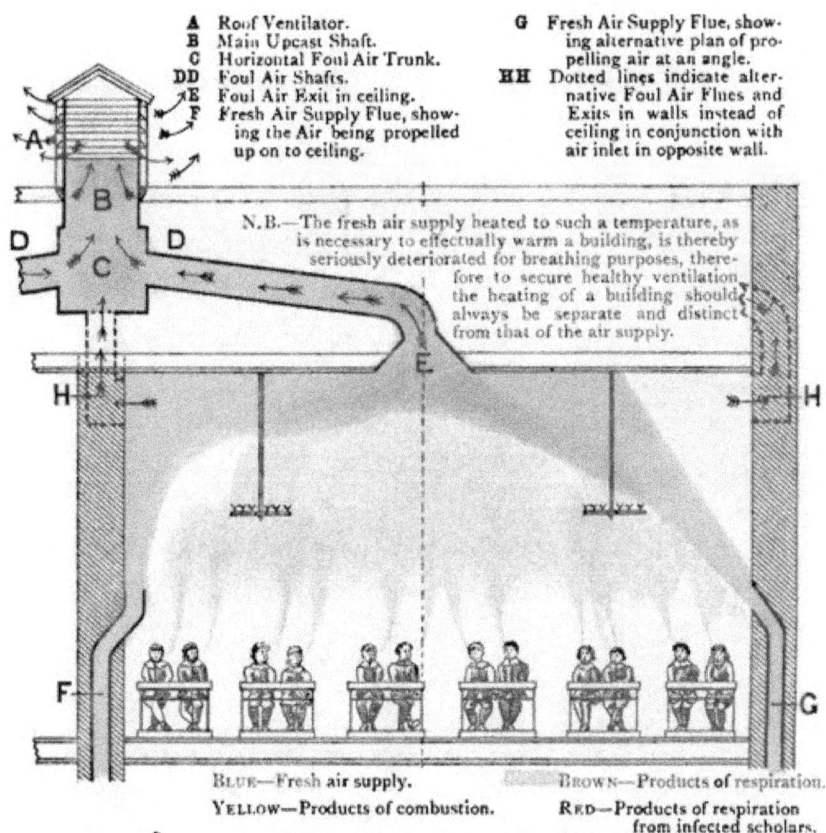

A Roof Ventilator.
B Main Upcast Shaft.
C Horizontal Foul Air Trunk.
DD Foul Air Shafts.
E Foul Air Exit in ceiling.
F Fresh Air Supply Flue, showing the Air being propelled up on to ceiling.

G Fresh Air Supply Flue, showing alternative plan of propelling air at an angle.
HH Dotted lines indicate alternative Foul Air Flues and Exits in walls instead of ceiling in conjunction with air inlet in opposite wall.

N.B.—The fresh air supply heated to such a temperature, as is necessary to effectually warm a building, is thereby seriously deteriorated for breathing purposes, therefore to secure healthy ventilation, the heating of a building should always be separate and distinct from that of the air supply.

BLUE—Fresh air supply.
YELLOW—Products of combustion.
BROWN—Products of respiration.
RED—Products of respiration from infected scholars.

"Mechanical ventilation by impulsion forces air into a building under pressure and at a high velocity—destructive of diffusion—causing disagreeable and dangerous draughts in the vicinity, and in the line of the inlets and outlets, the other parts of the building being left wholly unventilated, as the incoming columns of air usually travel, or are propelled, in a direct line to the nearest outlet, and there make their escape. Engine-driving columns of air through a building is not ventilating it. . . . It is wholly misleading to take the registered volumes of air forced into or out of a building at certain points as in any way indicating how often in a given time the whole of the air is changed."—*Building News.*

NOTE.—When a volume of air is forced in under pressure a certain portion is deflected and descends, as with the downward system, returning the respired air to be reinhaled, also the products of combustion."

Air Analysis and Defective Ventilation.

"It is now admitted that air analysis cannot always be accepted as correctly demonstrating the general purity of the air or efficiency of the ventilation, as in buildings where the ventilation has been condemned as dangerously defective the air analyses were seemingly all that could be desired, whilst the registered volumes of air forced in were enormous.

"Wonderful results may be shown in the way of tables of cubic feet of air propelled into a building, but they are illusory as in any way indicating the efficiency of the ventilation.

". . . . The expert further adds to his report [on the Ventilation of the Houses of Parliament] :—

" 'I never yet knew of a system of propulsion, pure and simple, that effected an efficient and satisfactory ventilation of any large building.'

"The Architect to the Local Government Board reports that 'mechanical ventilating arrangements are generally utter failures.'

"MECHANICAL VENTILATION ON THE DOWNDRAUGHT SYSTEM, BY IMPULSION, OR THE 'PLENUM' PRINCIPLE, APPLIED TO SCHOOLROOMS.

A   Roof Ventilator.
B   Main Upcast Shaft.
C   Horizontal Foul Air Trunk.
DD Connecting Pipes from Foul Air Flues.
EE Foul Air Flues.
FF Foul Air Exits into Flues EE.

GG Fresh Air Supply Flues through which the air is mechanically projected at a high velocity up on to the ceiling and then descends, escaping at Exits FF into Flues EE.

N.B.—The fresh air supply heated to such a temperature, as is necessary to effectually warm a building, is thereby seriously deteriorated for breathing purposes, therefore to secure healthy ventilation the heating of a building should always be separate and distinct from that of the air supply.

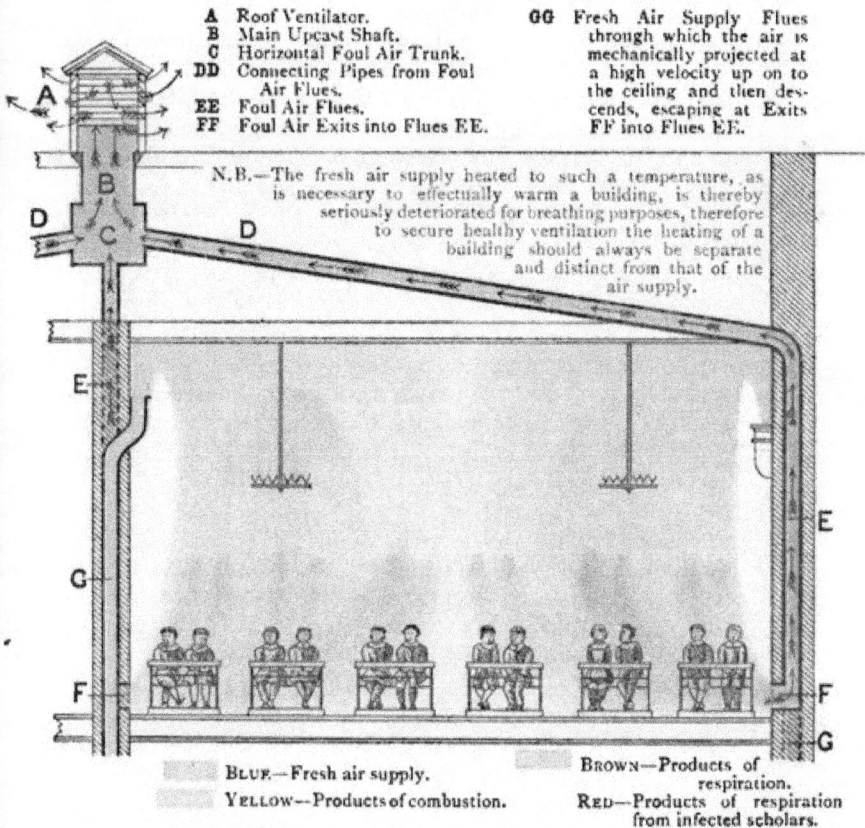

BLUE.—Fresh air supply.
YELLOW—Products of combustion.

BROWN—Products of respiration.
RED—Products of respiration from infected scholars.

"With all downdraught systems diluted foul air only is breathed, never fresh, pure air, as the respired and exhausted air, owing to its greater levity, ascends, and is returned by the downward current to the breathing level mixed with the product of combustion, and is there reinhaled. Only a very strong, and what would be an unendurable downdraught, could prevent the highly rarefied expired air from rising to a height sufficient to ensure its rebreathal. . . . It is universally condemned by scientific authorities as most pernicious, contrary to the laws of Nature, and—from causes which are well known—fatal to health, insidiously sowing the seeds of disease."—*Building News.*"

Over-heated Air a Danger to Health.

"One of the latest examples of the failure of mechanical ventilation was described by Mr. Thomas Blashill, the Architect to the London County Council, in a paper read before the Royal Institute of British Architects, on the new Council Chamber of the L.C.C., which is mechanically ventilated by a fan and fresh air warming apparatus. Mr. Blashill was 'not sure that these were of any use,' it being necessary at times to resort to natural ventilation in its most primitive form, and 'open doors, windows, coves, &c.,' both in warm and cold weather. When the ventilating apparatus was in operation, complaints were made of draughts, which Mr. Blashill attributed to descending currents.

"In the discussion which followed, Mr. Aston Webb, F.R.I.B.A., said, A building should be heated independently of the warmed fresh air supply, with which alone, according to not only his own experience, but also that of one of the highest authorities on the subject, Mr. Phipson, 'it was impossible to properly warm a chamber.' It would seem that the feeling of architects now tends towards keeping the two separate, as it has been found that warmed air employed for the combined purposes of heating and ventilation not only imperfectly warms the walls, but, owing to the high temperature to which the air is raised, its vitalising qualities necessary for healthy respiratory purposes are very considerably deteriorated.

"Scientists and sanitarians have long taught this, and we are glad to see that architects are now beginning to realise the importance of it."— *Local Government Journal.*

Printers: Sir Joseph Causton & Sons, Limited, London.

www.ingramcontent.com/pod-product-compliance
Lightning Source LLC
Chambersburg PA
CBHW021955190326
41519CB00009B/1266